T0093728

PACKAGING AND STORAGE OF FRUITS AND VEGETABLES

Emerging Trends

Postharvest Biology and Technology

PACKAGING AND STORAGE OF FRUITS AND VEGETABLES

Emerging Trends

Edited by

Tanweer Alam, PhD

First edition published 2022

Apple Academic Press Inc.
1265 Goldenrod Circle, NE,
Palm Bay, FL 32905 USA
4164 Lakeshore Road, Burlington,
ON, L7L 1A4 Canada

CRC Press
6000 Broken Sound Parkway NW,
Suite 300, Boca Raton, FL 33487-2742 USA
2 Park Square, Milton Park,
Abingdon, Oxon, OX14 4RN UK

© 2022 Apple Academic Press, Inc.

Apple Academic Press exclusively co-publishes with CRC Press, an imprint of Taylor & Francis Group, LLC

Reasonable efforts have been made to publish reliable data and information, but the authors, editors, and publisher cannot assume responsibility for the validity of all materials or the consequences of their use. The authors, editors, and publishers have attempted to trace the copyright holders of all material reproduced in this publication and apologize to copyright holders if permission to publish in this form has not been obtained. If any copyright material has not been acknowledged, please write and let us know so we may rectify in any future reprint.

Except as permitted under U.S. Copyright Law, no part of this book may be reprinted, reproduced, transmitted, or utilized in any form by any electronic, mechanical, or other means, now known or hereafter invented, including photocopying, microfilming, and recording, or in any information storage or retrieval system, without written permission from the publishers.

For permission to photocopy or use material electronically from this work, access www.copyright.com or contact the Copyright Clearance Center, Inc. (CCC), 222 Rosewood Drive, Danvers, MA 01923, 978-750-8400. For works that are not available on CCC please contact mpkbookspermissions@tandf.co.uk

Trademark notice: Product or corporate names may be trademarks or registered trademarks and are used only for identification and explanation without intent to infringe.

Library and Archives Canada Cataloguing in Publication

Title: Packaging and storage of fruits and vegetables : emerging trends / edited by Tanweer Alam, PhD.

Names: Alam, Tanweer, editor.

Series: Postharvest biology and technology book series.

Description: First edition. | Series statement: Postharvest biology and technology series | Includes bibliographical references and index.

Identifiers: Canadiana (print) 20210144351 | Canadiana (ebook) 20210144416 | ISBN 9781771889872 (hardcover) | ISBN 9781774638422 (softcover) | ISBN 9781003161165 (ebook)

Subjects: LCSH: Fruit—Packaging. | LCSH: Fruit—Storage. | LCSH: Vegetables—Packaging. | LCSH: Vegetables—Storage.

Classification: LCC TP440 .P33 2021 | DDC 664/.809—dc23

Library of Congress Cataloging-in-Publication Data

Names: Alam, Tanweer, editor.

Title: Packaging and storage of fruits and vegetables : emerging trends / edited by Tanweer Alam, PhD.

Description: First edition. | Palm Bay, FL : Apple Academic Press, 2021. | Series: Postharvest biology and technology | Includes bibliographical references and index. | Summary: "This new volume shares a plethora of valuable information on the recent advances in packaging and storage technologies used for quality preservation of fresh fruits and vegetables. This book, with chapters from eminent researchers in the field, covers several essential aspects of packaging and storage methods and techniques generally used in fruit and vegetables. Important considerations on selection and characteristics of packaging materials, recent packaging methods, storage hygiene and sanitation issues along with recent trends in storage technology are discussed in this volume. Key features: Provides an inclusive overview of fruit and vegetable requirements and available packaging materials and storage systems Imparts an understanding of the fundamentals of the impact of packaging on the evolution of quality and safety of fruits and vegetables Covers fundamental aspects of packaging and storage requirements, including mathematical modeling and mechanical and engineering properties of packaging materials Provides an in-depth discussion of innovative packaging and storage technologies, such as MA/CA packaging, active packaging, intelligent packaging, eco-friendly materials, etc., applied to fruit and vegetables Packaging and Storage of Fruits and Vegetables: Emerging Trends will be useful for graduate and postgraduate students and teaching professionals of horticultural science, food science and technology, packaging technology etc. It will also provide valuable scientific information to the academic scientific research community as well as to the packaging and storage industries for preservation of quality characteristics of fruits and vegetables. The professional community involved in handling processing and commercialization of horticultural crops will benefit as well"-- Provided by publisher.

Identifiers: LCCN 2021007369 (print) | LCCN 2021007370 (ebook) | ISBN 9781771889872 (hardcover) | ISBN 9781774638422 (paperback) | ISBN 9781003161165 (ebook)

Subjects: LCSH: Horticultural crops--Postharvest technology. | Fruit--Postharvest technology. | Vegetables--Postharvest technology.

Classification: LCC SB319.7 .P33 2021 (print) | LCC SB319.7 (ebook) | DDC 635/.04--dc23

LC record available at https://lccn.loc.gov/2021007369

LC ebook record available at https://lccn.loc.gov/2021007370

ISBN: 978-1-77188-987-2 (hbk)
ISBN: 978-1-77463-842-2 (pbk)
ISBN: 978-1-00316-116-5 (ebk)

ABOUT THE BOOK SERIES: POSTHARVEST BIOLOGY AND TECHNOLOGY

As we know, preserving the quality of fresh produce has long been a challenging task. In the past, several approaches were in use for the postharvest management of fresh produce, but due to continuous advancement in technology, the increased health consciousness of consumers, and environmental concerns, these approaches have been modified and enhanced to address these issues and concerns.

The Postharvest Biology and Technology series presents edited books that address many important aspects related to postharvest technology of fresh produce. The series presents existing and novel management systems that are in use today or that have great potential to maintain the postharvest quality of fresh produce in terms of microbiological safety, nutrition, and sensory quality.

The books are aimed at professionals, postharvest scientists, academicians researching postharvest problems, and graduate-level students. This series is intended to be a comprehensive venture that provides up-to-date scientific and technical information focusing on postharvest management for fresh produce.

Books in the series address the following themes:

- Nutritional composition and antioxidant properties of fresh produce
- Postharvest physiology and biochemistry
- Biotic and abiotic factors affecting maturity and quality
- Preharvest treatments affecting postharvest quality
- Maturity and harvesting issues
- Nondestructive quality assessment
- Physiological and biochemical changes during ripening
- Postharvest treatments and their effects on shelf life and quality
- Postharvest operations such as sorting, grading, ripening, de-greening, curing etc
- Storage and shelf-life studies
- Packaging, transportation, and marketing
- Vase life improvement of flowers and foliage
- Postharvest management of spice, medicinal, and plantation crops

- Fruit and vegetable processing waste/byproducts: management and utilization
- Postharvest diseases and physiological disorders
- Minimal processing of fruits and vegetables
- Quarantine and phytosanitory treatments for fresh produce
- Conventional and modern breeding approaches to improve the postharvest quality
- Biotechnological approaches to improve postharvest quality of horticultural crops

We are seeking editors to edit volumes in different postharvest areas for the series. Interested editors may also propose other relevant subjects within their field of expertise, which may not be mentioned in the list above. We can only publish a limited number of volumes each year, so if you are interested, please email your proposal wasim@appleacademicpress.com at your earliest convenience.

We look forward to hearing from you soon.

Editor-in-Chief:

Mohammed Wasim Siddiqui, PhD
Scientist-cum-Assistant Professor | Bihar Agricultural University
Department of Food Science and Technology | Sabour | Bhagalpur | Bihar | INDIA
AAP Acquisitions Editor, Horticultural Science
Founding/Managing Editor, Journal of Postharvest Technology

Email: wasim@appleacademicpress.com
wasim_serene@yahoo.com

Books in the Postharvest Biology and Technology Series:

Advances in Postharvest Technologies of Vegetable Crops
Editors: Bijendra Singh, PhD, Sudhir Singh, PhD, and Tanmay K. Koley, PhD

Emerging Technologies for Shelf-Life Enhancement of Fruits
Editors: Basharat Nabi Dar, PhD, and Shabir Ahmad Mir, PhD

Sensor-Based Quality Assessment Systems for Fruits and Vegetables
Editors: Bambang Kuswandi, PhD, and Mohammed Wasim Siddiqui, PhD

Emerging Postharvest Treatment of Fruits and Vegetables
Editors: Kalyan Barman, PhD, Swati Sharma, PhD, and
Mohammed Wasim Siddiqui, PhD

Innovative Packaging of Fruits and Vegetables: Strategies for Safety and Quality Maintenance
Editors: Mohammed Wasim Siddiqui, PhD, Mohammad Shafiur Rahman, PhD, and Ali Abas Wani, PhD

Insect Pests of Stored Grain: Biology, Behavior, and Management Strategies
Ranjeet Kumar, PhD

Packaging and Storage of Fruits and Vegetables: Emerging Trends
Editor: Tanweer Alam, PhD

Plant Food By-Products: Industrial Relevance for Food Additives and Nutraceuticals
Editors: J. Fernando Ayala-Zavala, PhD, Gustavo González-Aguilar, PhD, and Mohammed Wasim Siddiqui, PhD

Postharvest Biology and Technology of Horticultural Crops: Principles and Practices for Quality Maintenance
Editor: Mohammed Wasim Siddiqui, PhD

Postharvest Management of Horticultural Crops: Practices for Quality Preservation
Editor: Mohammed Wasim Siddiqui, PhD

ABOUT THE EDITOR

Tanweer Alam, PhD

Tanweer Alam, PhD, is the Director of the Indian Institute of Packaging (IIP), Mumbai, India. Before joining IIP, he worked as teaching faculty at Banaras Hindu University, Varanasi, SHUATS, Allahabad, and Jiwaji University, Gwalior, India. He has been working for more than a decade at different academic institutions and is also involved in the teaching of graduate, postgraduate, and doctoral students. His doctoral research work at the prestigious National Dairy Research Institute, Karnal, India, was based on modified atmosphere packaging of food products. He visited the University of Copenhagen, Denmark, under a faculty exchange program funded by the European Union under the East-West Food Project. He has participated in INTERPACK, Dusseldorf, Germany, and attended the award ceremony of WPO. He performed on the jury for Asia Star and World Star awards. Dr. Alam has organized several national and international conferences and workshops for promoting the packaging industry. He has published several research and technical articles in high-impact refereed journals and has delivered invited lectures both in India and abroad. He has also contributed more than 15 book chapters and also edited dozens of proceedings and technical souvenirs. He is an Associate Chief Editor of the *Asian Journal of Dairy Science* and *Journal of Postharvest Technology* and the Editor in Chief of the *Journal of Packaging Technology and Research* and is on the editorial board of the magazine *Processed Food Industry.* He is a life member of several scientific and academic associations.

CONTENTS

CONTRIBUTORS

Tanweer Alam
Indian Institute of Packaging, Mumbai, Maharashtra, India, E-mail: amtanweer@rediffmail.com

Z. R. A. A. Azad
Department of Postharvest Engineering and Technology, Aligarh Muslim University, Uttar Pradesh–202001, India

Soumitra Banerjee
Centre for Incubation, Innovation, Research, and Consultancy (CIIRC), Jyothy Institute of Technology, Bangalore, Karnataka, India

Aastha Bhardwaj
Department of Food Technology, Jamia Hamdard, New Delhi, India, E-mail: ab.0112@yahoo.com

Basharat Ahmad Bhat
Department of Life Sciences, Shiv Nadar University, Dadri, Uttar Pradesh–201314, India

Jinku Bora
Department of Food Technology, Jamia Hamdard, New Delhi–110062, India, E-mail: jinkubora@gmail.com

Nitin Chauhan
Shaheed Rajguru College of Applied Sciences for Women, University of Delhi, Delhi, India

Rajni Chopra
Institute of Home Economics, New Delhi, India

Gaurav Kr Deshwal
ICAR-National Dairy Research Institute, Karnal, Haryana, India, E-mail: ndri.gkd@gmail.com

Meenakshi Garg
Bhaskaracharya College of Applied Sciences, Delhi, India, E-mail: meenakshi.garg@bcas.du.ac.in

Pawas Goswami
Assistant Professor, Bhaskaracharya College of Applied Sciences, University of Delhi, Delhi, India, E-mail: pawasgoswami@gmail.com

Swarrna Haldar
Centre for Incubation, Innovation, Research, and Consultancy (CIIRC), Jyothy Institute of Technology, Bangalore, Karnataka, India

Entesar Hanan
Department of Food Technology, Jamia Hamdard, New Delhi–110062, India

Bisma Jan
Department of Food Technology, Jamia Hamdard, New Delhi, India

Yasmeena Jan
Department of Food Technology, Jamia Hamdard, New Delhi–110062, India

Meetaksh Kamboj
School of Agriculture and Food, The University of Melbourne, Parkville, Victoria–3010, Australia

Prabhjot Kaur
Shaheed Rajguru College of Applied Sciences, Delhi, India

Divya Kumari Kesharia
Centre of Food Science and Technology Institute of Agricultural Sciences, Banaras Hindu University, Varanasi, Uttar Pradesh, India

Priti Khemariya
Indian Institute of Packaging, Delhi, India

Kundan Kishore
Central Horticultural Experiment Station (ICAR-IIHR), Aiginia, Bhubaneswar–751019, Odisha, India

B. N. Skanda Kumar
Centre for Incubation, Innovation, Research, and Consultancy (CIIRC), Jyothy Institute of Technology, Bangalore, Karnataka, India

Muneeb Malik
Department of Food Technology, Jamia Hamdard, New Delhi–110062, India

Sahar Masud
Division of Livestock Production and Management, Sher-e-Kashmir University of Agricultural Sciences and Technology of Jammu, Jammu and Kashmir, India

Kamlesh Kumar Maurya
Centre of Food Science and Technology Institute of Agricultural Sciences, Banaras Hindu University, Varanasi, Uttar Pradesh, India

Sabyasachi Mishra
Assistant Professor, Department of Food Process Engineering, National Institute of Technology, Rourkela, Odisha 769008, India

Jayeeta Mitra
Agricultural and Food Engineering Department, Indian Institute of Technology Kharagpur, West Bengal, India, E-mail: jayeeta.mitra@agfe.iitkgp.ernet.ac.in

Narender Raju Panjagari
ICAR-National Dairy Research Institute, Karnal, Haryana, India

Rama Chandra Pradhan
Associate Professor, Department of Food Process Engineering, National Institute of Technology, Rourkela, Odisha 769008, India, E-mail: pradhanrc@nitrkl.ac.in

Sandeep Singh Rana
Assistant Professor, Department of Chemical Engineering, Vignan's Foundation Science Technology and Research, Vadlamudi, Andhra Pradesh 522213, India

Susmita Dey Sadhu
Bhaskaracharya College of Applied Sciences, Delhi, India

Deepa Samant
Central Horticultural Experiment Station (ICAR-IIHR), Aiginia, Bhubaneswar–751019, Odisha, India

Sadhana Sharma
National Institute of Food Technology and Entrepreneurship Management, Haryana, India

Arpit Shrivastava
Centre of Food Science and Technology Institute of Agricultural Sciences, Banaras Hindu University, Varanasi, Uttar Pradesh, India

Mohammed Wasim Siddiqui
Bihar Agricultural University, Sabour, Bihar, India

Swati Tiwari
Department of Food Science and Technology, Pondicherry University, Puducherry, Tamil Nadu, India

Thoithoi Tongbram
Department of Food Technology, Jamia Hamdard, New Delhi–110062, India

Abhishek Dutt Tripathi
Centre of Food Science and Technology Institute of Agricultural Sciences, Banaras Hindu University, Varanasi, Uttar Pradesh, India, E-mail: abhi_itbhu80@rediffmail.com

Mifftha Yaseen
Department of Food Technology, Jamia Hamdard, New Delhi–110062, India,
Mobile: +91-9953903217, E-mail: miffthayaseen@gmail.com

ABBREVIATIONS

ABA	abscisic acid
AMI	agricultural market information
APEDA	Agriculture and Processed Food Products Export Development Authority
BBC	British Broadcasting Corporation
BPYV	beet pseudo-yellows virus
C_2H_4	ethylene
CA	cellulose acetate
CA	controlled atmosphere
CAP Storage	cover and plinth storage
CAP	controlled atmosphere packaged
CFB	corrugated fiberboard
ClO_2	chlorine dioxide
CO_2	carbon dioxide
EMAP	equilibrium adapted atmosphere
EPS	expanded polystyrene
EU	European Union
EVOH	ethylene-vinyl alcohol
F&V	fruits and vegetables
FDA	Food and Drug Administration
Fe	iron
FEFO	first expire first out
FIFO	first in first out
FSSA	Food Safety and Standards Authority
FSSAI	Food Safety Standards Authority of India
GAC	granular activated carbon
GAs	gibberellins
GHP	good hygienic practices
GMPs	good manufacturing practices
GRAS	generally recognized as safe
GVA	gross value added
GWP	good warehousing practices
H_2O	water
HACCP	hazard analysis and critical control points

HCFC	hydrochlorofluorocarbon
HDPE	high-density polyethylene
HPP	high-pressure processing
IAA	indoleacetic acid
IPM	integrated pest management
$KMnO_4$	potassium permanganate
LAB	lactic acid bacteria
LDPE	linear density polyethylene
LDPE	low-density polyethylene
LIN	liquid nitrogen
LLDPE	linear low-density polyethylene
MA	modified atmosphere
MAP	modified atmosphere packaging
MCP	methyl cyclopropane
mm	millimeter
MOFPI	Ministry of Food Processing Industry
MT	million tons
N_2	nitrogen
O_2	oxygen
PA	polyamide
PAU	Punjab Agricultural University
PCOs	pest control operators
PCTFE	poly chloro trifluoro ethylene
Pd	palladium
PE	polyethylene
PET	polyethylene terephthalate
PG	polygalacturonase
PH	postharvest
PHA's	poly-hydroxy-alkanoates
PHB's	poly-hydroxy-butyrates
PLA	polylactic acid
PME	pectin methylesterase
PP	polypropylene
PPV	plum pox virus
PS	polystyrene
PVC	polyvinyl chloride
PVDC	polyvinylidene chloride
PVOH	polyvinyl alcohol
QACCP	quality analysis and critical control points

RFID	radiofrequency identification
RH	relative humidity
RSC	regular slotted container
SADH	succinic acid 2,2-dimethyl hydrazide
SCC	side-chain crystallizable
SO_2	sulfur dioxide
SPaV	strawberry pallidosis associated virus
SSOP'S	standard sanitation operating procedures
TICV	tomato infectious chlorosis virus
TiO_2	titanium dioxide
TMV	tobacco mosaic virus
ToCV	tomato chlorosis virus
ToMV	tomato mosaic virus
TPS	thermoplastic starch
TSWV	tomato spotted wilt virus
TTIs	time-temperature indicators
UAE	the United Arab Emirates
USA	the United States of America
USFDA	the United States Food and Drug Administration
UVRT	UV radiation technology
VP	vacuum packaging
WG	wheat gluten
WTO	World Trade Organization
WVTR	water vapor transmission rate
ZECC	zero-energy cool chamber

FOREWORD

Fruits and vegetables (F&V) are a rich source of protein, fatty acids, vitamins, minerals, dietary fibers, and antioxidants. They play a major role in maintaining good human health. Fruits and vegetables are highly perishable in nature and are greatly affected by several biotic and abiotic factors. A considerable amount (up to 35%) of fruits and vegetables are damaged due to poor postharvest practices, climatic conditions, distribution hazards, and microbial decay. Therefore, there is an urgent need for quality preservation of horticultural products in order to fulfill consumers' demand for healthy and safe foods. Improved postharvest handling and minimal processing along with storage in suitable packaging material can extend shelf life and maintain quality with reducing losses of horticultural products. High respiration rate, transpiration, ethylene production rate, and other metabolic processes are major criteria in the selection of suitable packaging material for storage and packaging of fresh produce.

This book appears to be the different of its kind on the storage and packaging technologies for quality preservation of fresh fruits and vegetables. The book contains the latest information on various aspects of storage and packaging for effective quality preservation of fruits and vegetables. It is hoped that this book will be greatly beneficial to graduate and postgraduate students, teaching professionals of horticultural science, food science technology, and packaging technology. The academic scientific research community and horticulture and packaging industry will also be benefitted.

—Prof. (Dr.) A. K. Srivastava

Chairman ASRB (Additional Charge),
Ministry of Agriculture and Farmer Welfare, GOI,
Krishi Anusandhan Bhawan-1, Pusa,
New Delhi–12, India

PREFACE

Fruit and vegetables play a significant role in human nutrition, especially as sources of protein, fatty acids, vitamins, minerals, dietary fibers, and antioxidants. A considerable amount of fresh fruits and vegetables (F&V) are damaged by several pre- and postharvest factors. Postharvest losses are estimated to range from 10% to 35% (fruits, 10% to 25%; vegetables, 15% to 35%) per year. Maintaining food safety and quality of horticultural products is very important for today's consumer's demand for healthy and safe food. Postharvest handling and processing technology maintains quality, reduces loss, and utilizes culls of agricultural products. In the present era, the reduction of losses in perishable products due to postharvest decay and damage has become a major objective of agricultural businesses.

Horticultural products are highly perishable in nature and are very easily affected by climatic conditions, distribution hazards, and microbial decay. In order to develop a suitable packaging material for horticultural produce, it is important to understand the biology of the produce since the high respiration rate, transpiration, ethylene production rate, and other metabolic processes create a special problem in the storage and packaging of fresh produce. Temperature, relative humidity (RH), and ventilation also play a very important role in determining the postharvest life of fresh produce. Thus, the packaging material for fresh produce is chosen to provide protection against bruising and physical injury, microbial contamination, and deterioration; moisture/weight loss; slow down respiration rate, to delay ripening; increase storage life; and control ethylene concentrations in the package.

This book is one of the volumes of the book series *Postharvest Biology and Technology.* This volume covers prevent the recent advances in packaging and storage technologies used for the quality preservation of fresh fruits and vegetables. This book is comprised of 12 chapters written by the eminent professionals in the field. The book covers several essential aspects of packaging and storage technologies generally used in fruit and vegetables. Important considerations on the selection and characteristics of packaging materials, different recent packaging methods, and storage hygiene and sanitation issues, along with recent trends in storage technology, have been given in this book.

The book will be useful for graduate and postgraduate students and teaching professionals of horticultural science, food science, and technology, packaging technology, etc. It will also impart valuable scientific information to the academic scientific research community as well as to the packaging and storage industries for the preservation of quality characteristics of fruits and vegetables. The professional community involved in handling the processing and commercialization of horticultural crops will be benefitted as well.

We express our gratitude to each and every author who has taken an interest in writing the chapters and, by this way, always kept us enthusiastic to bring the book in the final shape.

Our appreciation also goes for Apple Academic Press for readily accepting the book chapters and for the excellent end product. Without their active support, the book could not have been completed.

CHAPTER 1

PHYSICOCHEMICAL AND PHYSIOLOGICAL CHANGES DURING STORAGE

SANDEEP SINGH RANA,[1] RAMA CHANDRA PRADHAN,[2] and SABYASACHI MISHRA[2]

[1]*Assistant Professor, Department of Chemical Engineering, Vignan's Foundation Science Technology and Research, Vadlamudi, Andhra Pradesh 522213, India*

[2]*Associate Professor, Department of Food Process Engineering, National Institute of Technology, Rourkela, Odisha 769008, India, E-mail: pradhanrc@nitrkl.ac.in (R. C. Pradhan)*

ABSTRACT

The history of storing good products is as old as human civilization. Season variations in the harvest need storage of stocks in excess of immediate requirement to adjust the mismatch between supply and demand. The quality can be maintained over a period of time by proper storage. Relevant literature based upon long accumulated experience indicates that chemical changes in food grains are a natural process and occur continuously, resulting in the loss/gain of weight, the changing of physiological appearance, the loss of nutritional/food value, the loss of culinary properties, and ultimately the total destruction of a food product. Stock deterioration is mainly by post-harvest processing and storage conditions. Conditions in storage such as humidity, temperature, presence, and attack by insects, rodents, fungus, etc., determine to a large extent the safe storage life of stock. Of all parameters of chemical changes, moisture is the most important factor. The process of breakdown of triglycerides, resulting in the production of free fatty acids value and breakdown of starch into smaller fraction s resulting in increase in reducing sugar add to the process of deterioration. Auto-oxidation of unsaturated fatty acids, producing peroxide, and growth of fungus has been recognized as the major parameters in changing the chemical composition. Starting from the domestic method of storing food, there is a wide range of storing system befitting to the requirement of consumer. Modern storage facilities, whether they are meant for bulk or small-scale storage, have to satisfy a number of basic requirements. The summarized requirement of storage facility is to protect the commodity from excessive temperature, moisture, and from attacks of rodents.

1.1 INTRODUCTION

There are many factors which influences the quality of product while storage. Few of these factors can be enlisted as product condition during and before storage, type of storage structure, time or duration of storage (storage period), and type of product handling while storage (Neethirajan et al., 2007). The presence of foreign matter like dust, sand from field, chaff, etc., influences the quality of stored product and need to be managed specifically.

The factor which is considered to have a major influence while storage of produce is the initial quality of produce itself. Other factors which influence the quality are moisture content of product, relative humidity (RH) of product, temperature, reparation, and heat production (Multon, 1989). These factors are discussed in subsections.

1.1.1 MOISTURE

While storage of product, the moisture content inside the product comes in equilibrium with the air inside the storage structure and the air within the produce. This will develop the most favorable condition for the growth of microorganism, which will harm product quality with time. Hence, it is always advised to dry the produce to safe storage moisture content before storing it for long durations.

1.1.2 RELATIVE HUMIDITY (RH)

Different insects, pests, microorganism, and rodents need different levels of humidity and temperature for their favorable growth. For example, insects can easily grow if RH is between 30 and 50%. Must mites need 60% RH for their favorable development. Whereas in the case of bacteria and molds, the level of RH are 90% and 70%, respectively (Dyson, 1999). However, RH is not the only criteria for development of spoilage organisms in produce. It is better to quote that combine action of RH, moisture content and temperature can be responsible for development of harmful organisms in produce (Emekci et al., 1989; Gbolade and Adebayo, 1993; Kundu et al., 2005).

1.1.3 TEMPERATURE

There are maximum chances that stored grains are high temperatures inside the storage structure on a hot day because of unaerated grain bulks and storage of for many months and the insulation properties of grains. This high temperature can easily cause heat damage to the grains. Similar type of heat damage is also common in fruits and vegetables (F&V) due to their high respiration rate. The temperature also influence the biological and enzymatic activities inside the produce; hence it also influence the rate of spoilage of produce. The problem like moisture migration, decolorization, dense air, are the result of inappropriate temperature while storage.

1.1.4 RESPIRATION AND HEAT PRODUCTION

Respiration is a very common and essential phenomenon in all the living cells. Respiration is in the presence of oxygen is known as aerobic respiration. Aerobic respiration causes the breakdown of food, i.e., fats, carbohydrates, proteins, or vitamins in living organisms. The energy released during respiration produces heat. Although the stored grain are having very low respiration rate but if the moisture content of grains are high then respiration rate increases drastically which cause increase of respiration heat and results in increase of grain temperature.

The changes which occur during the storage of produce are discussed in this chapter. The change inside grains varies from changes occurs in fruits and vegetable during storage. All those changes are discussed in details and different sections.

1.2 QUALITY CHANGES IN STORED FOOD GRAINS

Foodgrains are mostly stored in bulk storages or in silos. During the storage of food grains, they undergo various changes in their physical, chemical, and biological attributes. These changes occur because of enzymes, biochemical, and enzymes produce due to insects, pests or microbes, or many other factors. All those factors cause degrading of quality of grains during storage.

The constituent which plays the major role during storage of food grains is moisture content. High moisture content in food grains is considered as a hazard for storage and transportation because it encourages pests, microbes, and insects (White and Jayas, 2003; Weinberg et al., 2008; Manickavasagan et al., 2008a). Even mites become more frequent troublemaker in food grain storage with high moisture content, and mites become a very serious problem in the storage of grains, especially in winters. It was found that at normal storage condition that is at ordinary temperature and humidity, different species of insects, pests, and microbes need different time to develop. Vast knowledge and complete history of pests or insects are required to trace the origin of infestation accurately.

The need for bulk storage arose from various situations. The most common is a rapid increase in demand from consumers. Since long-term storage system ideally requires a low capital investment in the storage structure but a high degree of security from losses during storage (Manickavasagan et al., 2008b).

1.2.1 PHYSICAL CHANGE

The fresh, sound, and healthy grains are having good luster, hardness, and shiny. The various physical changes in the grains undergo during storage are dullness in color, bores in grains, sprouting of seeds, damaged kernels due to bad weather conditions.

1.2.1.1 COLOR CHANGE

Discolorization of grains caused mainly by heat produced because of respiration of grains and insects as well as storage temperature. The discolorization starts from the germ area (cotyledon) and it continues through sides and back of the kernel (endosperm) in a continuous band. Provided the band continues, there is no minimum width requirement (White and Jayas, 2003). It is generally considered that heating of grain is mostly started because of respiration and sometimes it initiated by insects, but in recent studies, it was found that the end products of the use of the grain as food by the insects have been shown responsible. These end products are heat, water, and carbon dioxide. Heat and water are well known to be important factors for decolorization of produce during storage (Figure 1.1).

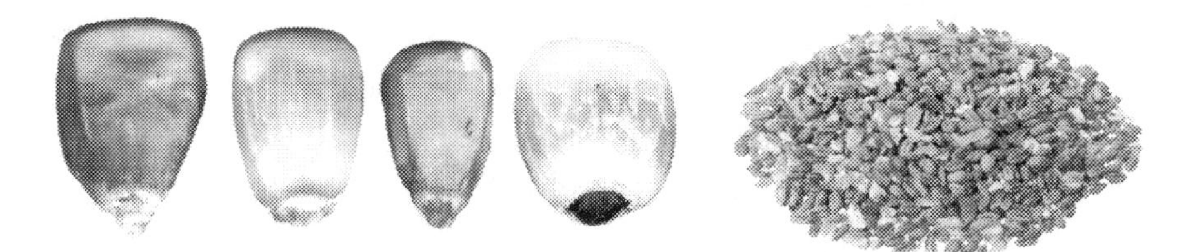

FIGURE 1.1 Discoloration in corn due to heat.

White (1995) reported the changes of color of various rice, stored at a different temperature, caused by the non-enzymatic. Further, browning accompanied by losses of reducing sugars, amino nitrogen, and free amino acids also occurred.

1.2.1.2 BORE IN GRAIN

Bore in grain during storage is found as most common during storage (Navarro et al., 1984; Navarro, 2006; Phillips and Throne, 2010). The quantity of grain eaten by the insects is not always as important as the spoilage of the grain using factors secondary to the feeding of the insects and the capacity risk to close by grain through insect reproduction and unfold (Figure 1.2).

FIGURE 1.2 A lesser grain borer on a wheat kernel.

Credit: Clemson University-USDA Cooperative Extension Slide Series, Bugwood.org.

Insects and pest problems are very much difficult to resolve in the progressive or undeveloped country, because of unavailability of equipment to maintain cleanliness and favorable conditions for safe storage which must safeguard against pests and insects. In general cleanliness is not considered as important as it should be during storage. Cleanliness of a grain storage structures, silos or bins involves the disposal of spoiled or infected grains, waste from rodents, pests, insects, or other harmful foreign impurities such as dust and chaff can accumulate, especially in places that are inclined to be warm and damp.

1.2.1.3 SPROUTING OF SEEDS

Storage of grain for 3–4 months after harvest results in sprouting or aging effect or ripening may occur due to their physicochemical nature (Martin et al., 1917; Maneepun, 2003; Collins et al., 2005). Five to six enzymes will be available even after the aging of grains, which is due to the accumulation of substrate receptors, except cytochrome C reductase. Aside from degradation, interaction among membrane-bound particulate substances, such as starch in compound granules, phytin in globoids in aleurone protein bodies, and storage protein, glutelin, and prolamin in protein bodies during sprouting. Starch protein interaction, specially glutelin starch interaction decreasing with sprouting (Figure 1.3).

1.2.1.4 DAMAGED KERNELS

The kernels of grains got damaged due to various factors. Major factors like improper storage conditions, moisture in stored produce, Uneven heating of grains, insects, or pests. Rotted kernels, sawfly kernels, ruptured kernels midge damaged kernels are the few common damages causes in kernels (Figure 1.4) (Deshpande et al., 1982; Kells et al., 2001; Jayas and White, 2003).

Greely (1978) reported that Sawfly damaged Kernels are shriveled or distorted. Whereas, Rotted kernels are swollen, spongy (under pressure), discolored, and soft because of decomposition by bacteria or fungi. Ruptured Kernels are considered to be damaged when the split in the cheek extends at least half the length of the cheek or if both cheeks are split to any degree. Midge damaged kernels are with ruptured bran from back or at side, there will a white line on kernel and kernel will be distorted (Antunes and Sgarbrieri, 1979; Aidoo, 1993).

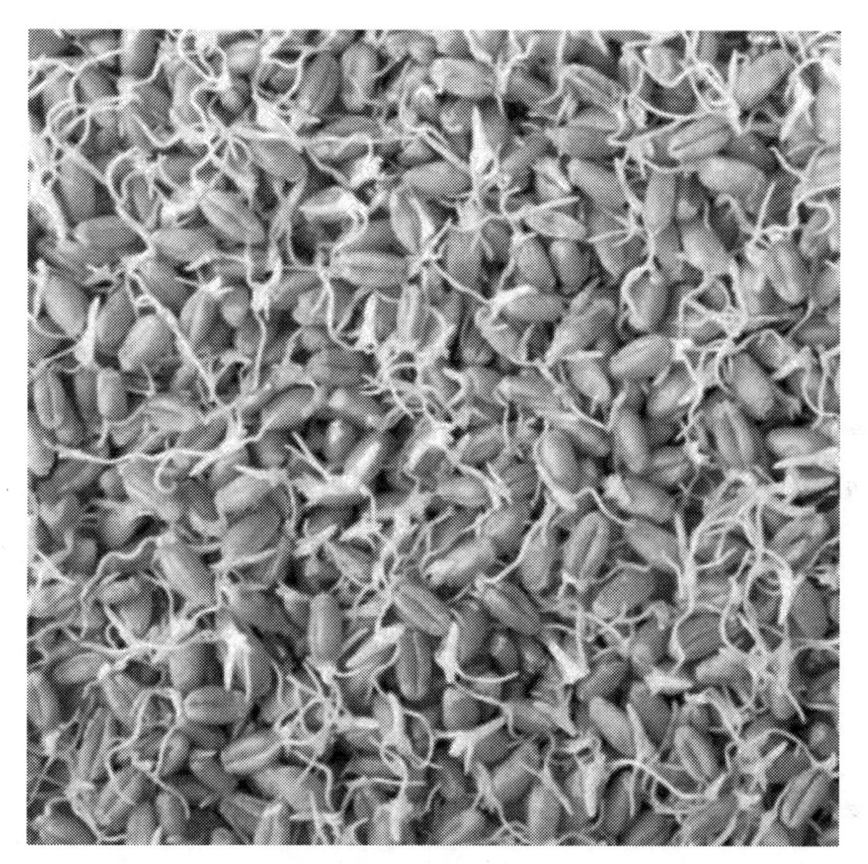

FIGURE 1.3 Sprouting in wheat grains during storage.

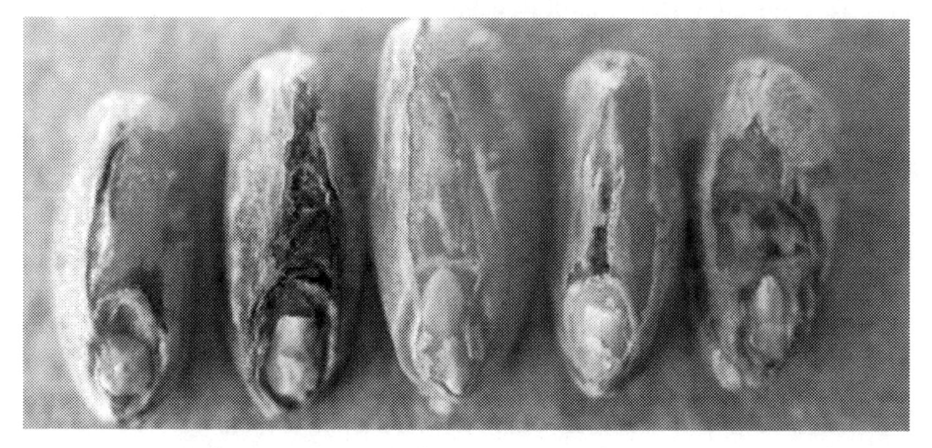

FIGURE 1.4 Damaged kernel of wheat during storage.

Agricultural Grains are also known as a biological organism, which deteriorates with time during storage (Cox and Collins, 2002). Maintaining grains in suitable storage conditions will protect grain and will also help to maintain its optimum quality. Weatherproofing grain of storage spaces or structures can effectively keep out insects. These structures can also be effective against keeping moisture away from grain, which is major causes for damaged kernels.

1.2.2 CHEMICAL CHANGES

The chemical changes in food grains are a natural process and occur continuously resulting in the loss/gain of weight, changing of physical appearance, loss of nutritional/food value, loss of culinary properties, and ultimately, total destruction of the grain (Stanley and Aguilera, 1985; Tipples, 1995; White, 2003). The process is influenced by a number of factors, namely post-harvest processing, and storage conditions such as humidity, temperature, attack of insects, rodents, fungus, etc.

1.2.2.1 CHANGE IN THE COMPOSITION

Adler et al. (2000) has reported the effect of storage on the chemical composition. The amylase activity in freshly harvested rice decreased rapidly; there was an improvement in the cooking quality. There was an increase in acidity and the peroxide value of the fat present in the different samples of rice, in increase being at a maximum for husked rice.

1.2.2.1.1 Carbohydrates

Solutions of amylose and starch isolated from fresh rice have a slightly higher specific viscosity than the corresponding constituents from old rice. All vitamins in cereals degrade during storage, particularly at high moisture and temperature (Varnava, 2002; Yousif et al., 2007). Water-soluble vitamins degrade slowly during storage. In addition, riboflavin and pyridoxine are sensitive to light. Thiamine content of brown rice reducers during parboiling but diffuses to the endosperm during the process. Vitamin A in maize is also light sensitive and also degrades during storage. Vitamin E and alpha-tocopherol also degrade during storage of maize (Kon and Sanshuck, 1981; Liu et al., 1992).

1.2.2.1.2 Protein

Tuan and Phillips (1991); Xia and Sun (2002); Nayak et al. (2005) also reported a dramatic increase in paste viscosity during storage. Attempts to explain these functional changes have focused on the properties of rice components such as starch, protein, and lipids and the interactions among them during storage.

1.2.2.1.3 Lipid

According to Esper and Miihlbauer (1998), oxidation of lipids, especially the unsaturated fatty acids results in typical rancid flavor, odor, and taste. Hydrolysis of lipids also increases free fatty acid contents, which is considered as a sensitive index for grain deterioration (Henderson and Henderson, 1968; Gowen et al., 1993; Kendall and Pimentel, 1994; Islam et al., 2009).

Pabis et al. (1998) reported the changes in lipid composition in wheat during storage deterioration. Singh (1999) reported the deteriorative changes in the oil fraction of stored parboiled rice. Again Overhults et al. (1973) reported that the total extractable and organic acids decreased during maturation but increased during storage; more strongly as moisture or temperature increased. They observed that grain deterioration was accompanied by the lowering of polar lipids.

Uma and Pushpamma (1986) observed the increase in oleic acid and that linoleic acid decreased with storage time. The same trend was observed in fraction containing free fatty acids and mono- and diglyceride.

1.2.2.2 CONTAMINANTS

Foodgrains are usually contaminated with foreign materials viz. stones, chaffs, poisonous weeds, excreta of insects, pests, rodents, etc., which gives poor look to the grains. Limits of weed presence, uric acid and insect excreta should be as described by the Government of India (FCI) for the stored food grains. Before any recommendation is made in this regard, complete evaluation of these materials is required with respect to their chemical composition, toxins, residues of insecticides, pesticides, and finally the *in vivo* feeding value of different categories of livestock of grain infected by molds in the 6th month of storage (Page, 1949; Menkov, 2000; Sousa et al., 2008).

Analysis conducted on the grains indicates a decreasing percentage of moisture, and increasing acid and peroxide values with increased age (time period between stock entry and sampling date). The fungal load of the sampled grains varies widely between samples taken from various government storage units. This variation is most probably attributable to the extent of contamination imparted to the grains during post-harvest handling, processing, and storage (Bell, 1978; Garcia and Stanley, 1989; Lewis, 1921; Mertz and Yao, 1993).

1.2.2.3 TOXIC COMPOUNDS GENERATION

With low moisture grain, fungi are not consideration and the insect populations appear significantly stressed that populations, irrespective of their initial density, maybe reducing to light infections only. Typically there will be a sharp initial drop in oxygen and a rise in carbon dioxide content, the extend of which depends on the level of infection and moisture content of grains. The changes are then reversed, presumably from suppression of the infestation, and it depends on the degree of sealing of grains during storage. The alternative solution to reduce infestation of grain during storage is disinfestations of grains before storage.

Moisture content is also the reason for the fluctuation of toxins in stored grains. According to Mathioulakis et al. (1998), aflatoxin content in the stored grain for 1 year was noted maximum from grain stored in a room type structure followed by bulk covered while the minimum value of aflatoxin was observed in straw-clay bin. Whereas Tiwari et al. (2010) reported an incident of fungal flora determination of aflatoxin in rice.

Barrozo et al. (2004) reported preservation of rough rice by cold storage and observed that yeast count fluctuated markedly and viability of the higher-moisture rice decreased markedly. They reported on average quite high loads of fungal growth in both un-husked rice and husked rice samples collected from various government and private storage depots and markets. The moisture content has been found to influence favorably the higher loads of fungi. However, the storage period has little influence on the fungal content when the moisture content of the stored samples remains below 16%.

1.2.3 BIOCHEMICAL AND BIOLOGICAL EFFECTS

The major aroma principle in cooked aromatic rice 2-acetyl 1-pyrroline decreases in milled rice during storage. Ruengssakulrach et al. (1989)

studied the influence of storage on the microbial population of rough rice. He reported that rice with 18 to 20% moisture content, sealed in gas-lined bin for 7 months became sour. Some molds survived but did not increase; the facultative anaerobic organisms markedly increased, and yeast increased tremendously in most of the layers of the piles of rice. Other biochemical and biological hazard in stored grains are insects and weevils, which often and destructive. Which leaves their excreta inside the stored grains and cause hazardous microclimate inside the bins or godowns.

1.3 QUALITY CHANGES IN STORED FRUITS AND VEGETABLES (F&V)

F&V need to be protected from post-harvest losses to ensure the demands of consumers and processors. Improper handling and storage facilities drastically increase the wastage of food resources. The main reasons for wastage of food resources are heavy spoilage after harvest in fruits and weight loss, shriveling, and early senescence in vegetables. The problems have been multiplied due to the seasonal availability and localized production. Various post-harvest have been evolved to increase the shelf life of fruit and vegetables. Common among these treatments are low-temperature storage, controlled atmospheric storage, hypobaric storage, high humidity storage, and treatments with plant growth hormones like ethylene or heat treatments (Jayas and Jeyamkondan, 2002). The success of these treatments depends upon their suitability simplicity and economic feasibility (Figure 1.5).

1.3.1 PHYSICAL CHANGE

1.3.1.1 COLOR CHANGES

Color change in F&V during storage is served as visual maturity index. The main pigments in F&V are carotenoids and chlorophyll.

Chlorophyll dominates during the unripe stage, but while storage ripening is another process occurs simultaneously, so it causes a change of color of fruit and vegetables during storage. Change in color is majorly due to degradation of chlorophyll and reason in degradation of chlorophyll is chlorophyllase activity and also its synthesis (Fields et al., 2001).

Carotenoids are a group of yellow pigments present in green fruit or vegetable tissues before maturation and dominate in ripe fruit. Accumulation of carotenoids is common in many F&V during storage. Carotenoids differ in their structure in composition from one species to another. Simple forms are lycopene and beta-carotene but available in other highly complex forms.

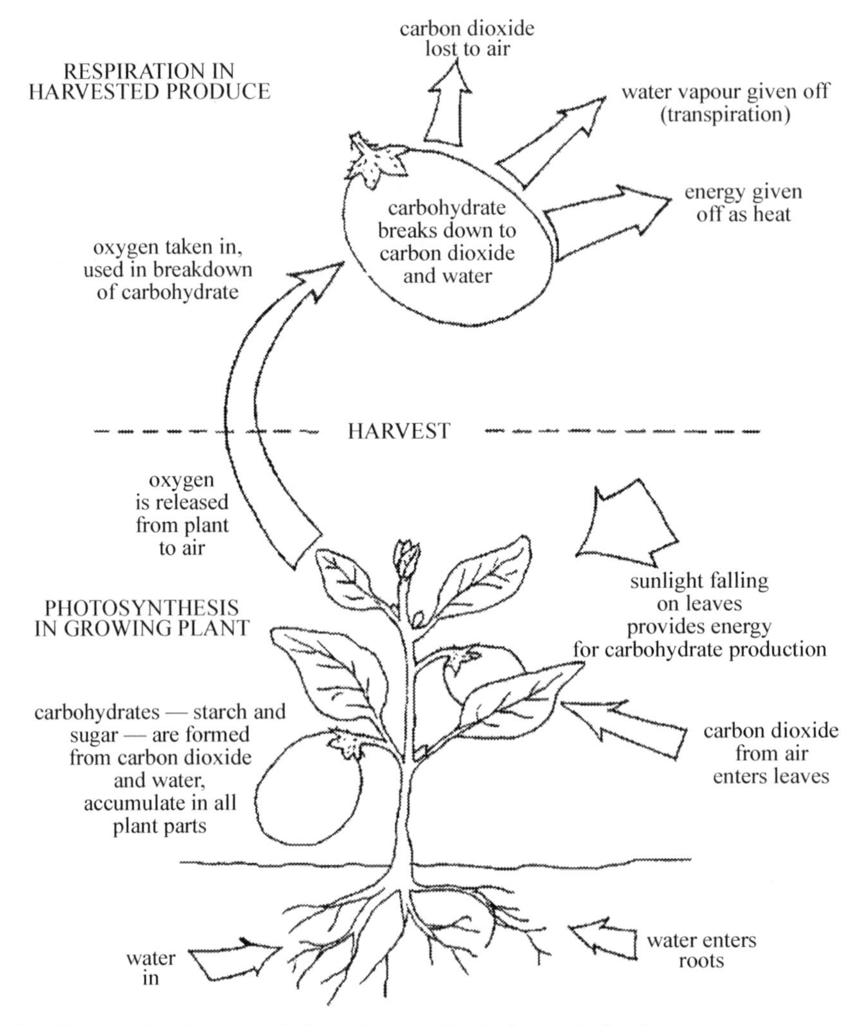

FIGURE 1.5 Physiology of fruits and vegetables before and after harvesting.

Source: Reprinted from Prevention of post-harvest food losses: fruits, vegetables and root crops. © 1989 FAO. http://www.fao.org/3/t0073e/T0073E02.htm

Anthocyanins are water-soluble; phenolic pigments are known to be stored in vacuoles. Six principal anthocyanidins occur in fruits in three glycosides that impart a red color to apple varieties, strawberries, grapes, etc., (Fields and

White, 2002; Emekci et al., 2004). These pigments may interact with other phenols such as tannins and some flavonoids resulting in discolorization.

1.3.1.1.1 Factors Affecting Color Change

Several internal and external physicochemical factors influence color change because pigment change brought about by an array of complex biochemical reactions:

1. **Light:** It delays loss of chlorophyll. According to Harborne et al. (1971), loss of chlorophyll was enhanced by red light in harvested potatoes wherein lycopene, and beta-carotene formation was induced. The effect can be nullified by far-red and dark treatments.
2. **Temperature:** Low temperatures cause pigment change. Low temperature (15°C can effectively utilized in degreening citrus fruits without ethylene (Kostyukovsky et al., 2002). Temperature 25°C is optimum for biosynthesis of lycopene. Above 30°C favored the accumulation of lycopene but not carotenes.
3. **Hormone Changes:** Auxins, cytokinins, and gibberellins (GAs) occur in low quantities in maturing fruits during storage. Auxins are known for the delay in fruit maturation which is very helpful to keep the fruit and vegetables safe for long storages. The oxidative turnover of auxins is the onset of ripening.

Cytokinins and GAs are known to delay maturity in many F&V. They interfere with chlorophyll degradation and biosynthesis of carotenoids and anthocyanins (Garcia et al., 1998; Dyson, 1999). Abscisic acid (ABA) and succinic acid 2,2-dimethyl hydrazide (SADH) enhanced the color of various fruit and vegetables. Their action is similar to auxins.

1.3.1.2 TEXTURAL CHANGE

The texture is a most important attribute, and it very difficult to determine and measure the texture of stored fruit and vegetable. The texture of stored fruit and vegetable changes with time. Different F&V have their specific texture type, and it also changes with storage time and ripening. Texture profile of any sample is greatly influenced by parameters like maturity, the structure of cell wall, and composition of cell walls. In large parenchymatous tissues of fruit cells, the degree of contact between cells is also an added factor.

Pectins, cellulose, and other polygalacturonase (PG) are the main components in cell wall carbohydrates. Pectin makes up one-third of primary dry matter in the cell wall of F&V. Along with pectin, hemicelluloses contribute to the bulk of non-cellulosic dry material of primary cell wall and are considered to exhibit micro-dispersity. Cellulose is present in the primary cell wall in the linear association of the polymer molecules called fibrils.

In addition to these complex carbohydrates, 0.5–2% of the dry weight of fruits and vegetable cell walls is constituted by proteins. Lignin is absent or present in very low concentrations in the cells of F&V at their edible stages. It appears in senescent vegetables in their secondary cell walls. It is a complex polymer derived from phenolic compounds.

The common sensory texture testing method are squeezing in hand, bending between the teeth, the subjective sensory methods for measuring textural properties of food with relation to human perception is well-reviewed by El-Nahal et al. (1989).

1.3.1.3 STRUCTURAL CHANGES

The visible storage changes are the result of histological and histochemical changes in underlying tissues. Although all other facets of storage have been extensively investigated, the structural changes have received very little attention. A comprehensive account of anatomical structures of plant tissues, including F&V have been published by Hincks and Stanley (1986). The use of electron microscopy is an impetus for the study of ultrastructural changes.

1.3.1.3.1 Surface Changes

SEM provides a characteristic three-dimensional view of the material and hence widely used to study the surfaces of F&V. Hentges et al. (1991) studies the fine structure of cuticle on fruit surface of different apple cultivars. In ripe apple, the structure of cuticle is homogenous and evenly spread on the surface, often interspersed with pores and transcuticular canals. However, in *Vaccinium elliotti* the structure of surface waxes in raw, immature fruits varied from flat to upright platelets either horizontally or vertically oriented. These variations disappeared with the advance of maturity and ripening. At the edible ripe stage, only a limited amount of platelets or waxes were retained with a complete absence of rodlets. Hayakawa and Breeene (1982) showed changes in the permeability of calcium penetration in apples at various developmental stages.

The change in the surface waxes of cherries at various stages of storage was observed by SEM and LM. Cracking of cuticle at ripe stages of berry was correlated with the water adsorption. This could overcome by ethyl oleate treatment which helps in redistribution of waxes (Bodnaryk et al., 1999). The total epicuticular waxes increased continuously throughout fruit development (Jayas and Jeyamkondan, 2002).

1.3.1.3.2 Anatomical Changes

Inter and intracellular structural changes of F&V during storage have been studied by SEM and TEM. The softening of fruit has been evident by the changes in the degree of cell separation, cell wall thickness, and starch hydrolysis in some cases. Earlier microscopic studies on different F&V revealed that the parenchymatous cell wall becomes extremely thin and tenuous during storage Brooker et al. (1974). However, in avocadoes cell walls lost their shape and structure without any alteration in thickness. Dissolution of middle lamella and the gradual disintegration of fibrillary material throughout the cell wall are responsible for cell separation in apples, pears, and tomatoes (Figure 1.6).

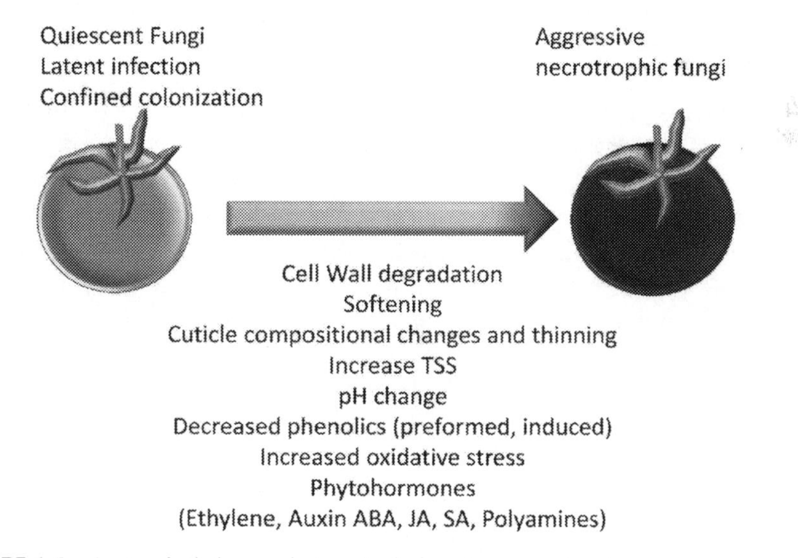

FIGURE 1.6 Anatomical changes in tomato during storage or ripening.

Comparative account of structural changes in carrots during growth, storage, and disease was recorded along with various fixing methods.

Influence of processing on the microstructure of carrot and beans were studied by Hayakawa and Breeene (1982).

Starch granules which are abundantly present in most of the unripe F&V were well studied for structural disintegration during ripening and storage. Straited structures on the surface of starch granules of banana were observed by Hentges et al. (1991). Formation of numerous pinholes on the surface of starch granules by the action of amylase was also visible. Internally these pores show a terraced appearance. The starch granules, however, become fragile and broken into small irregular fragments in beans after storing them for ten days.

1.3.1.4 CHANGE IN AROMA AND FLAVOR

Fruits and vegetables are important sources of proteins, fibers, vitamins, and minerals. It is equally important for their esthetic qualities such as aroma and color. Flavor of F&V are the complex mixtures of many high molecular weight volatile compounds earlier; sensory evaluation was the only technique to determine the quality of aroma and odor in food samples. The advert of Gas chromatography and various sophisticated instruments let to the characterization and evaluation of various chemical compounds. The methodology of extraction and determination of various volatiles compounds were reviewed by many researchers. The role of amino acids and proteins in food flavors were reviewed by Bell (1978); Garcia and Stanley (1989); Lewis (1921); Mertz and Yao (1993).

The primary aroma and flavor developments occur during ripening and storage while the secondary aroma and flavor components are developed during, chewing crushing or processing. Various pre and post-harvest factors responsible for the manifestation of aroma in F&V. Aroma development is very low during unripe and pre-climacteric fruit and starts with the onset of climacteric and develops fully after the climacteric rise. Volatiles dissolved in non-cellular cuticle is recorded in may fruit due to their solubility in polar solvents (Garcia and Stanley, 1989; Lewis, 1921).

1.3.2 CHEMICAL CHANGE

1.3.2.1 RIPENING

Fruits soften while vegetables toughen during later stages of storages. Degradation of cellulytic and pectic material results in textural weakening of tissues. Moreover, damage to the semi-permeability of the cell membrane leading to loss of turgor and softening. A reduction in cell wall pectin due to solubility was reported in the whole range of fruits. Change in poly provides been intensively investigated by Fields et al. (2001). In addition to a net loss of non-cellulosic neutral sugar, residues have been reported by Bell (1978); Garcia and Stanley (1989); Lewis (1921); Mertz and Yao (1993) (Figure 1.7).

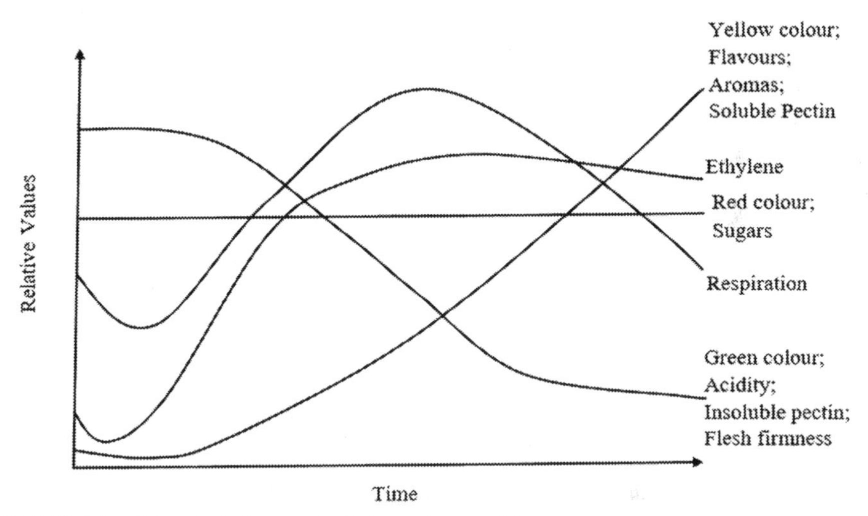

FIGURE 1.7 Chemical changes in fruits and vegetables.

These changes are brought about by an array of carbohydrates degrading enzymes of which PG, pectin methylesterase (PME), and cellulose are the most important. Their activity is low or absent in unripe F&V, and it increases with the ripening of F&V. PG is cell wall-bound and brings about solubilization of pectin. PME is also cell wall-bounded, and its basic role is to modify pectin structure before PG action. Moreover, hence it promotes PG activity and softening of tissues.

1.3.2.2 CHANGE IN THE COMPOSITION

1.3.2.2.1 Carbohydrate Metabolism

Carbohydrates are the principal source of energy and main respiratory substrate in F&V. A marked reduction in starch content with an increase in sugars with respiratory rise was noted in F&V (Jayas and Jeyamkondan, 2002). These changes were well reflected by increased activity of glycolytic enzymes, phosphofructokinase, and pyruvate kinase which promote glycolysis during climacteric (Garcia et al., 1998; Dyson, 1999). In addition, activation or de novo synthesis of hydrolytic enzymes like alpha-amylase and beta-amylase was suggested (Fields et al., 2001). Garcia et al. (1998) were found degradation of cell wall carbohydrates (pectin and cellulose) by their respective degrading enzymes.

1.3.2.2.2 Effect on TCA

No major changes were observed in the functional capacity of mitochondria at various stages of climacteric. Pyruvate levels as a substrate for oxidative activities increased with storage Tiwari et al. (2010). In banana, TCA cycle increased with storage result in a decrease in pH. In apple decrease in malic acid was observed (Jayas and Jeyamkondan, 2002).

1.3.2.2.3 Protein and Nucleic Acid

Utilization of generated ATP during respiration or storage for protein synthesis is less elucidated. There is a rise in protein synthesis during pre-climacteric and disease attack during storage of F&V (Figure 1.8).

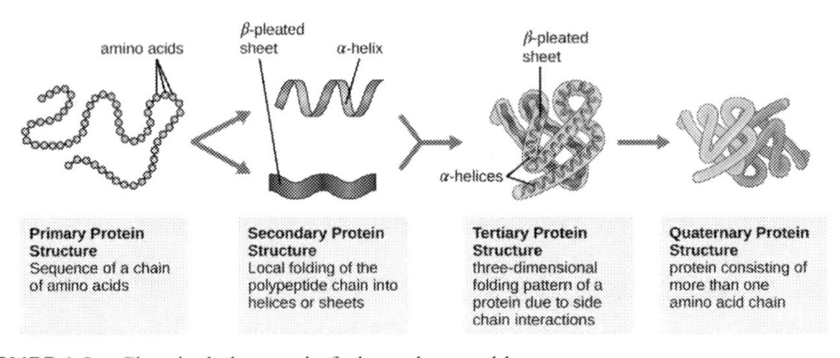

FIGURE 1.8 Chemical changes in fruits and vegetables.

Source: National Human Genome Research Institute.

1.3.2.2.4 Cellular Organism

Increase in membrane permeability during storage is a common phenomenon. An increase in permeability is recorded just before to climacteric rise in many F&V. Subsequently enhanced the potassium ion leakage in cytosolic concentration which is a requirement for PK and positive modulator of PFK, is expected to enhance storage.

In non-climacteric F&V, the storage changes are the result of biochemical aging unlike perfective changes in climacteric fruit.

1.3.2.3 RESPIRATION

Respiration is an essential and irreversible active metabolic process of fruit and vegetables. This is associated with cataclysmic, physiological, histochemical, and biochemical changes. Respiration is a catabolic and anabolic process. Various physicochemical and biological changes during respiration are well-reviewed (Figure 1.9).

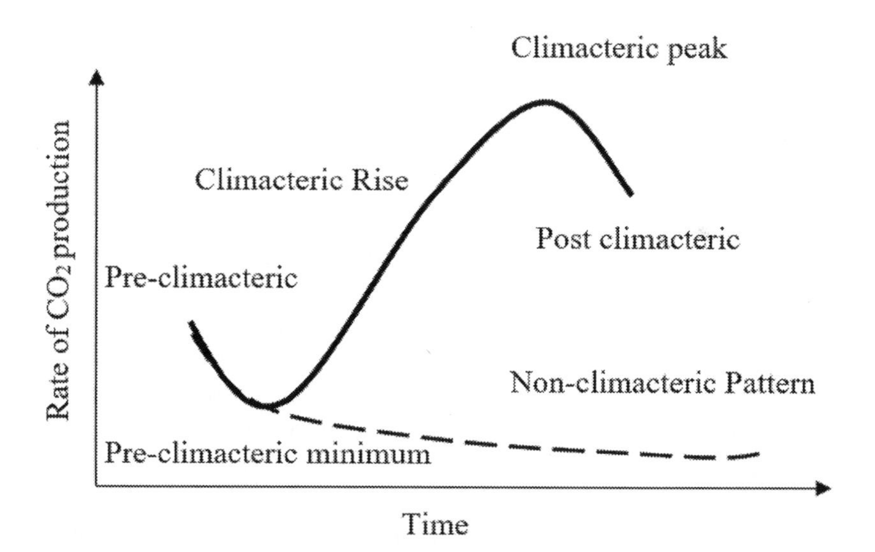

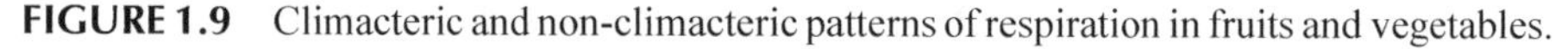

FIGURE 1.9 Climacteric and non-climacteric patterns of respiration in fruits and vegetables.

Two major types of respiration were observed in F&V:

i. Climacteric respiration; and
ii. Non-Climacteric Respiration.

Climacteric manifests the climax of all metabolic activities ultimately heralding the onset of senescence. The major metabolic activities related to respiration are:

1. **Endogenous Ethylene Generation:** All climacteric F&V produce the ethylene endogenously. It biosynthesis the various ripening changes during storage. Ethylene starts respiratory CO_2 generation and ripening of F&V. Traces of trapped ethylene in intercellular space of tissues in early stages of storage were detected without any rise in respiration rate (Table 1.1).

TABLE 1.1 Classification of Fruits and Vegetables to Their Maximum Ethylene Production Rate

Ethylene Production Rate	Fruits
Very low (0.01–0.1)	Cherry, citrus, grapes, pomegranate, strawberry
Low (0.1–1)	Blueberry, kiwifruit, peppers, persimmon, pineapple, raspberry
Moderate (1–10)	Banana, fig, honeydew melon, mango, tomato
High (10–100)	Apple, apricot, avocado, cantaloupe, nectarine, papaya, pear, and plum
Very high (>100)	Cherimoya, mamey apple, passion fruit, sapota

1.4 CONCLUSION

From all the parameters of chemical changes in food grains during storage, changes in moisture content are the most important factor. Breakdown of triglycerides, resulting in the production of free fatty acids and acid value and the breakdown of starch into smaller fractions result in an increase of reducing sugar. Autoxidation of unsaturated fatty acids, producing peroxide, and growth of fungus have been recognized as other major parameters. It may be noted that the change in protein occurs very slowly and thus may be ignored for practical purposes.

Once grain becomes infected, it should be kept separate from clean lots until it has been treated properly to free it of insects by fumigation or by exposure to low or high temperatures or low moisture conditions that kill the pests. It is well known that dry malt with a moisture content of about 3% never becomes infested with weevils. The species of insects that attack sound grain are relatively few and cannot do any appreciable amount of damage in milled products, but a number of insects follow after those that attack sound grain. These secondary insects feed upon the hulls left by the others and may cause the grain to heat. They also work in milled products and are a serious pest in mills.

KEYWORDS

- **chlorophyll**
- **decolorization**
- **microorganism**
- **moisture content**
- **pectin methylesterase**
- **polygalacturonase**
- **succinic acid 2,2-dimethyl hydrazide**

REFERENCES

Adler, C., Corinth, H. G., & Reichmuth, C., (2000). Modified atmospheres. In: Subramanyam, B. H., & Hagstrum, D. W., (eds.), *Alternatives to Pesticides in Stored-Product IPM* (Vol. 5, pp. 105–146). Kluwer Academic Publishing, Norwell, MA.

Aidoo, K. E., (1993). Post-harvest storage and preservation of tropical crops. *Int. Biodeterior. Biodegradation, 32*, 161–173.

Antunes, P. L., & Sgarbrieri, V. C., (1979). Influence of time and conditions of storage on technological and nutritional properties of a dry bean (*Phaseolus vulgaris* L.) variety rosinha G2. *J. Food Sci., 44*, 1703–1706.

Barrozo, M. A. S., Sartori, D. J. M., & Freire, J. T., (2004). A study of the statistical discrimination of the drying kinetics equations. *Food Bioprod. Process, 82*, 219–225.

Bell, E. A., (1978). Toxins in seeds. In: Harborne, J. B., (ed.), *Biochemical Aspects of Plant and Animal Coevolution* (pp. 143–161). Academic Press, New York.

Bodnaryk, R., Fields, P., Xie, Y., & Fulcher, K., (1999). *Insecticidal Factors from Field Pea.* US Patent: 5,955,082.

Brooker, D. B., Bakker-Arkema, F. W., & Hall, C. W., (1974). *Drying Cereal Grains.* AVI, Westport.

Collins, P. J., Daglish, G. J., Pavic, H., & Kopittke, R. A., (2005). Response of mixed-age cultures of phosphine-resistant and susceptible strains of lesser grain borer, *Rhyzopertha Dominica*, to phosphine at a range of concentrations and exposure periods. *J. Stored Prod. Res., 41*, 373–385.

Cox, P. D., & Collins, L. E., (2002). Factors affecting the behavior of beetle pests in stored grain, with particular reference to the development of lures. *J. Stored Prod. Res., 38*, 95–115.

Deshpande, S. S., Sathe, S. K., Salunkhe, D. K., & Cornforth, D. P., (1982). Effects of dehulling on phytic acid, polyphenols and enzyme inhibitors of dry beans (*Phaseolus vulgaris*). *J. Food Sci., 47*, 1846–1850.

Dyson, T., (1999). *World Food Trends and Prospects to 2025* (Vol. 96, pp. 1–13). Proceedings of the National Academy of Sciences, USA.

El-Nahal, A. K. M., Schmidt, G. H., & Risha, E. M., (1989). Vapors of *Acorus calamus* oil: A space treatment for stored product insects. *J. Stored Prod. Res., 25*, 211–216.

Emekci, M., Navarro, S., Donahaye, E., Rindner, M., & Azrieli, A., (2002). Respiration of *Tribolium castaneum* (Herbst) at reduced oxygen concentrations. *J. Stored Prod. Res., 38*, 413–425.

Emekci, M., Navarro, S., Donahaye, E., Rindner, M., & Azrieli, A., (2004). Respiration of *Rhyzopertha Dominica* (F.) at reduced oxygen concentrations. *J. Stored Prod. Res., 40*, 27–38.

Esper, A., & Miihlbauer, W., (1998). Solar drying: An effective means of food preservation. *Renewable Energy, 15*, 95–100.

Fields, P. G., & White, N. D. G., (2002). Alternatives to methyl bromide treatments for stored-product and quarantine insects. *Ann. Rev. Entomol., 47*, 331–359.

Fields, P. G., Xie, Y. S., & Hou, X., (2001). Repellent effect of pea (*Pisum sativum*) fractions against stored-product insects. *J. Stored Prod. Res., 37*, 359–370.

Garcia, E., Filisetti, T. M. C. C., Udaeta, J. E. M., & Lajolo, F. M., (1998). Hard-to-cook beans (*Phaseolus vulgaris*): Involvement of phenolic compounds and ectates. *J. Agric. Food Chem., 46*, 2110–2116.

Garcia-Vela, L. A., & Stanley, D. W., (1989). Protein denaturation and starch gelatinization in hard-to-cook beans. *J. Food Sci., 5495*, 1284–1292.

Gbolade, A. A., & Adebayo, T. A., (1993). Fumigant effects of some volatile oils on fecundity and adult emergence of *Callosobruchus maculatus F. Insect Sci. Applic., 14*, 631–636.

Gowen, A. A., Tiwari, B. K., Cullen, P. J., McDonnell, K., & O'Donnell, C. P., (2010). Applications of thermal imaging in food quality and safety assessment. *Trends Food Sci. Technol., 21*, 190–200.

Greeley, M., (1978). Appropriate rural technology: Recent Indian experience with farm-level food grain storage research. *Food Policy, 3*(1), 39–49.

Harborne, J. B., Boulter, D., & Turner, B. L., (1971). *Chemotaxonomy of the Leguminosae*. Academic Press, London.

Hayakawa, I., & Breene, W. M., (1982). A study on the relationship between cooking properties of adzuki bean and storage conditions. *J. Fac. Agr. -Kyushu Univ., 27*, 83–88.

Henderson, J. M., & Henderson, S. M., (1968). A computational procedure for deep-bed drying analysis. *J. Agric. Eng. Res., 13*, 87–95.

Hentges, D. L., Weaver, C. M., & Nielsen, S. S., (1991). Changes of selected physical and chemical components in the development of the hard-to-cook bean defect. *J. Food Sci., 56*, 436–442.

Hincks, M. J., & Stanley, D. W., (1986). Multiple mechanisms of bean hardening. *Food Technol., 21*, 731–750.

Islam, M. S., Hasan, M. M., Xiong, W., Zhang, S. C., & Lei, C. L., (2009). Fumigant and repellent activities of essential oil from *Coriandrum sativum* (L.) (Apiaceae) against red flour beetle *Tribolium castaneum* (Herbst) (Coleoptera: Tenebrionidae). *J. Pest Sci., 82*, 171–177.

Jayas, D. S., & Jeyamkondan, S., (2002). PH-post-harvest technology: Modified atmosphere storage of grains meats fruits and vegetables. *Biosyst. Eng., 82*, 235–251.

Jayas, D. S., & White, N. D. G., (2003). Storage and drying of grain in Canada: Low-cost approaches. *Food Control, 14*, 255–261.

Kells, S. A., Mason, L. J., Maier, D. E., & Woloshuk, C. P., (2001). Efficacy and fumigation characteristics of ozone in stored maize. *J. Stored Prod. Res., 37*, 371–382.

Kendall, H. W., & Pimentel, D., (1994). Constraints on the expansion of the global food supply. *Ambio, 23*, 98–205.

Kon, S., & Sanshuck, D. W., (1981). Phytate content and its effect on cooking quality of beans. *J. Food Process Preserv., 5*, 169–178.

Kostyukovsky, M., Rafaeli, A., Gileadi, C., Demchenko, N., & Shaaya, E., (2002). Activation of octopaminergic receptors by essential oil constituents isolated from aromatic plants: Possible mode of activity against insect pests. *Pest Management Science, 58*, 1–6.

Kundu, K. M., Das, R., Datta, A. B., & Chatterjee, P. K., (2005). On the analysis of drying process. *Drying Technol., 23*, 1093–1105.

Lewis, W. K., (1921). The rate of drying of solid materials. *Ind. Eng. Chem., 13*, 427.

Liu, K., McWatters, K. H., & Phillips, R. D., (1992). Protein in solubilization and thermal destabilization during storage as related to hard-to cook defect in cowpeas. *J. Agric. Food Chem., 40*, 2483–2487.

Maneepun, S., (2003). In: Shanmugasundaram, S., (ed.), *Traditional Processing and Utilization of Legumes*. Report of the APO Seminar on Srisuma, N., Hammerschmidt, R., Uebersax, M. A.,

Manickavasagan, A., Jayas, D. S., & White, N. D. G., (2008a). Thermal imaging to detect infestation by *Cryptolestes ferrugineus* inside wheat kernels. *J. Stored Prod. Res., 44*, 186–192.

Manickavasagan, A., Jayas, D. S., White, N. D. G., & Paliwal, J., (2005). Applications of thermal imaging in agriculture: A review. *Can. Soc. Eng. Agric. Food Biol. Syst.,* Paper no. 05–002.

Manickavasagan, A., Jayas, D. S., White, N. D. G., & Paliwal, J., (2008b). Wheat class identification using thermal imaging: A potential innovative technique. *Trans. ASABE, 51*, 649–651.

Martin-Cabrejas, M. A., Esteban, R. M., Perez, P., Maina, G., & Waldron, K. W., (1997). Changes in physicochemical properties of dry beans (*Phaseolus vulgaris* L.) during long-term storage. *J. Agric. Food Chem., 45*, 3223–3227.

Mathioulakis, E., Karathanos, V. T., & Belessiotis, V. G., (1998). Simulation of air movement in a dryer by computational fluid dynamics: Application for the drying of fruits. *J. Food Eng., 36*, 182–200.

Menkov, N. D., (2000). Moisture sorption isotherms of chickpea seeds at several temperatures. *J. Food Eng., 45,* 189–194.

Mertz, F. P., & Yao, R. C., (1993). *Amycolatopsis alba* sp. nov., isolated from soil. *International Journal of Systematic Bacteriology, 43*(4), 715–720.

Multon, J. L., (1989). Spoilage mechanisms of grains and seeds in the post-harvest ecosystem, the resulting losses, and strategies for the defense of stocks. In: Multon, J. L., (ed.), *Preservation and Storage of Grains, Seeds and Their by-Products Cereals, Oilseeds, Pulses and Animal Feed* (pp. 3–11). CBS Publishers, New Delhi.

Navarro, S., (2006). Modified atmospheres for the control of stored-product insects and mites. In: Heaps, J. W., (ed.), *Insect Management for Food Storage and Processing* (2nd edn., pp. 105–146). AACC International, St. Paul, MN.

Navarro, S., Donahaye, E., Kashanchi, Y., Pisarev, V., & Bulbul, O., (1984). Airtight storage of wheat in PVC covered bunker. In: Ripp, B. E. et al., (eds.), *Controlled Atmosphere and Fumigation in Grain Storages* (pp. 601–614). Amsterdam, Elsevier.

Nayak, M. K., Daglish, G. J., & Byrne, V. S., (2005). The effectiveness of Spinosad as a grain protectant against resistant beetle *and psocid pests of stored grain in Australia. Journal of Stored Products Research, 41*(4), 455–467.

Neethirajan, S., Karunakaran, C., Jayas, D. S., & White, N. D. G., (2007). Detection techniques for stored-product insects in grain. *Food Control, 18*, 157–162.

Overhults, D. G., White, G. M., Hamilton, H. E., & Ross, I. J., (1973). Drying soybeans with heated air. *Trans. ASAE, 16*, 112–113.

Pabis, S., Jayas, D. S., & Cenkowski, S., (1998). *Grain Drying: Theory and Practice*. John Wiley, New York.

Page, G. E., (1949). *Factors Influencing the Maximum Rates of Air-Drying Shelled Corn in Thin-Layer*. Purdue University, West Lafayette.

Phillips, T. W., & Throne, J. E., (2010). Biorational approaches to managing stored-product insects. *Annual Review of Entomology, 55*, 375–397.

Phillips, T. W., Cogan, P. M., & Fadamiro, H. Y., (2000). Pheromones. In: Subramanyam, B. H., & Hagstrum, D. W., (eds.), *Alternatives to Pesticides in Stored-Product IPM* (pp. 273–302). Kluwer Academic Publishers, Boston.

Ruengssakulrach, S., Bennink, M. R., & Hosfield, G. L., (1989). Storage induced changes of phenolic acids and the development of hard-to-cook in dry beans (*Phaseolus vulgaris*, var. Seafarer). *J. Food Sci., 54*, 311–318.

Singh, U., (1999). Cooking quality of pulses. *J. Food Sci. Technol. Mysore, 36*, 1–14.

Sousa, A. H., Faroni, L. R. D. A., Guedes, R. N. C., Tótola, M. R., & Urruchi, W. I., (2008). Ozone as a management alternative against phosphine resistant insect pests of stored products. *J. Stored Prod. Res., 44*, 379–385.

Stanley, D. W., & Aguilera, J. M., (1985). A review of textural defects in cooked reconstituted legumes-the influence of structure and composition. *J. Food Biochem., 9*, 277–323.

Tipples, K. H., (1995). Quality and nutritional changes in stored grain. In: Jayas, D. S., White, N. D. G., & Muir, W. E., (eds.), *Stored Grain Ecosystems* (pp. 325–351). Marcel Dekker, New York.

Tiwari, B. K., Brennan, C. S., Curran, T., Gallagher, E., Cullen, P. J., & O' Donnell, C. P., (2010). Application of ozone in grain processing. *J. Cereal Sci., 51*, 248–255.

Tuan, Y. H., & Phillips, R. D., (1991). Effect of hard-to-cook defect and processing on the protein and starch digestibility of cowpea. *Cereal Chem., 68*, 413–419.

Uma-Reddy, M., & Pushpamma, P., (1986). Effect of storage on amino acid and biological quality of protein in different varieties of pigeon pea, green gram, and chickpea. *Nutr. Rep. Int., 33*, 1020–1029.

Varnava, A., (2002). Hermetic storage of grain in Cyprus. In: Batchelor, T. A., & Bolivar, J. M., (eds.), *Proc. Int. Conf. n Alternatives to Methyl Bromide* (pp. 163–168). Sevilla, Spain.

Weinberg, Z. G., Yan, Y., Chen, Y., Finkelman, S., Ashbell, G., & Navarro, S., (2008). The effect of moisture level on high moisture maize (*Zea mays* L.) under hermetic storage conditions-in vitro studies. *Journal of Stored Products Research, 44*, 136–144.

White, N. D. G., & Jayas, D. S., (2003). Quality changes in grain under controlled atmosphere storage. In: Navarro, S., & Donahaye, E., (eds.), *Proceedings of the International Conference on Controlled Atmosphere and Fumigation in Grain Storages* (pp. 205–214). Caspit Press Limited, Jerusalem, Israel.

White, N. D. G., (1995). Insects, mites, and insecticides in stored-grain ecosystems. In: Jayas, D. S., White, N. D. G., & Muir, W. E., (eds.), *Stored-Grain Ecosystems* (pp. 123–167). Marcel Dekker, New York.

Xia, B., & Sun, D. W., (2002). Application of computational fluid dynamics (CFD) in the food industry: A review. *Comput. Electron. Agric., 34*, 5–24.

Yousif, A. M., Kato, J., & Deeth, H. C., (2007). Effect of storage on the biochemical structure and processing quality of adzuki bean (*Vigna angularis*). Processing and Utilization of Legumes, Japan, 9–14 October 2000 Asian Productivity Organization, Tokyo. *Food Rev. Int., 23*, 1–33, 53–62.

CHAPTER 2

QUALITY CHARACTERISTICS OF PACKAGING MATERIALS AND CONTAINERS USED FOR STORAGE OF FRESH PRODUCE

AASTHA BHARDWAJ,[1] BISMA JAN,[1] and TANWEER ALAM[2]

[1]*Department of Food Technology, Jamia Hamdard, New Delhi, India, E-mail: ab.0112@yahoo.com (A. Bhardwaj)*

[2]*Indian Institute of Packaging, Mumbai, Maharashtra, India*

ABSTRACT

The use of appropriate packaging materials for fresh horticultural produce is quint essential to retain/improve shelf life and keeping quality of the products as well as to minimize post harvest losses throughout the supply chain. With the optimum temperature, mixture of gases and water activity, an efficient control over the shelf life for fruit and vegetables can be maintained. These include all forms of consumer or unit packaging, transport packaging and unit load packaging. Gunny bags, bamboo baskets, corrugated fiberboard (CFB) boxes, metal trunks, etc., are a few of the commercially utilized variants. Perforated PE bags and net bags are generally used for retail and wholesale purposes. In addition, sophisticated technologies such as MAP and Intelligent packaging have been demonstrated to add value and credibility to the packaging supply chain. This chapter describes in detail, the various types of conventional and modern packaging materials and techniques employed for fresh produce, their quality characteristics, uses, and limitations.

2.1 INTRODUCTION

Fruits and vegetables (F&V), being a good source of nutrients such as minerals, vitamins, phytochemicals, and fiber, have a great significance in healthy human nutrition. The horticulture industry consists of fruits, vegetables, as well as ornamental flowers. With changing habits, mindset, and demographics, people are also getting more aware about proper food habits. The consumption of F&V has increased to a great extent owing to hectic and stressful lifestyle. China tops the current list of fruit production 154.364 million tons (MT) followed by India (82.631 MT), Brazil (37.774 MT), USA (26.986 MT) and Spain (17.699 MT). India is the second-largest producer of fruits as well as vegetables having a global share of approximately 10% and 15%, respectively. In spite of all the advances in preservation technologies, almost 20 to 30% of the total horticultural production is lost and wasted every year due to major postharvest losses and lack of competent infrastructure. F&V are respiring commodities that undergo ripening and aging throughout their lifespan. Sugars are broken down by consuming oxygen and producing carbon dioxide, water, and heat. This adversely affects the product's freshness, taste, and nutritional quality. In addition, these foods liberate ethylene, in small quantities, that accelerate the ripening process in F&V. This biochemical reaction is dependent on temperature. Low temperature allows slow respiration and less ethylene liberation. Furthermore, fruits and vegetables contain a high amount of water content ranging from 75–95%. Relative humidity (RH) of such perishables is as high as 98%. Under normal atmospheric conditions, fruits, and vegetables result in rapid drying, which causes wilting and shriveling due to loss of rigidity and shrinkage of the cells. The use of most favorable packaging material together with optimum storage temperature conditions contribute to delaying of maturation and senescence processes in respiring fruit and vegetables. The purpose of utilizing proper packaging materials for fresh horticultural produce is to help the consumers choose optimum packaging for their products to improve shelf life and quality of the product and to reduce losses in the supply chain. With the optimum temperature, mixture of gases and water activity, an efficient control over the shelf life for fruit and vegetables can be maintained.

Packaging for fresh horticultural produce can be classified on the basis of their distribution system for which it is primarily intended, namely, consumer or unit packaging, transport packaging, and unit load packaging. For efficient handling, transporting, and dispersal of the fresh produce, bags, crates, cartons, bulk bins, and palette containers are among the most widely used convenient containers. In addition, these cater to the majority of diverse requirements of wholesalers, retailers, and consumers as well as processing operators. In addition to these, tissue paper wraps, trays, cups or pads, may be used as supplementary packaging materials to reduce abrasion damage. Small packages with relatively few layers of produce, particularly delicate products, e.g., kiwi, passion fruit, etc., are utilized to reduce compression damages. In order to avoid abrasion damages during transportation and to physically separate the individual produce units, molded trays may also be used. F&V may also be wrapped individually in a tissue or wax-based paper to facilitate physical protection and reduce microbiological infestation inside the pack. Gunny bags, bamboo baskets, corrugated fiberboard (CFB) boxes, metal trunks, etc., are also available in different capacities and shapes. Perforated PE bags and net bags are generally used for retail as well as wholesale purposes for F&V commercially (Figure 2.1).

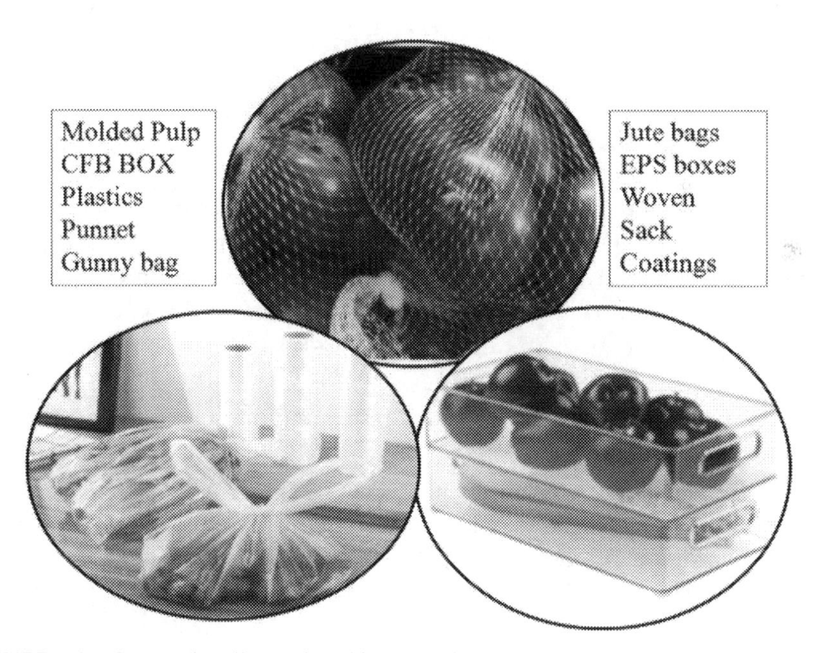

FIGURE 2.1 Conventionally used packing materials for fruits and vegetables.

For instance, molded pulp trays and CFB boxes are used for citrus fruits like oranges, tangerines, and mandarins. 3-Ply CFB boxes (regular slotted container (RSC), expanded polystyrene (EPS) boxes for custard apple; 3-Ply or 5-Ply CFB boxes with ventilation holes of varying capacity for green chili, brinjal, bottle guard, etc., are commonly used packaging materials in the markets globally.

The major important parameters for extending the storage life of horticultural produce are temperature, moisture, RH, and combination of various atmospheric gases such as oxygen, carbon dioxide, and ethylene. The senescence and aging of the perishable produce can be significantly brought down if the aforementioned parameters are taken care off. Packaging undoubtedly provides a safe edge over not only physical damages, but also chemical as well as microbiological defects too and thus providing an optimal shelf life to the F&V (Thompson, 1996). This chapter describes in detail the various types of conventional and modern packaging materials and techniques employed for fresh produce, their quality characteristics, uses, and limitations.

2.2 TYPES OF PACKAGING MATERIALS AND THEIR CHARACTERISTICS

2.2.1 PAPER AND MESH BAGS

This kind of package includes all bags other than plastic film bags. These include fiber net bags, plastic net bags, paper bags, and plastic mesh bags. A paper bag, generally made up of brown kraft paper, is a traditional form of packaging for carrying fresh fruit and vegetables. Mesh and net bags are used to pack potatoes, onions, turnips, cabbage, and other citrus fruits. Bags are usually closed by twist ties or cellophane adhesives. In addition to its low cost, mesh has the advantage of uninhibited airflow as onions in particular, require good ventilation (Mandal, 2015). Mesh bags are generally preferred by the supermarket managers because they make produce visually attractive and appealing to inspire purchase. However, mesh bags as well as paper bags provide less protection against chemical, physical, and biological contaminants (Figure 2.2).

2.2.2 WOOD

Wood-based rigid packaging materials such as planks, crates, pallets, and pallet bins, are a portable platform on which goods are stored or moved. These may be reusable, allow adequate ventilation to the produce, and facilitate stacking of the produce during storage and transportation. In order to provide additional cushioning properties, these can be woven outside using a coir rope in the form of wire-bound veneers. In addition, wooden wire-bound crates are sturdy and rigid forms of wood-based packaging used for beans, sweet corn, and other commodities requiring hydro cooling. Their limitations include their heavyweight, pointed edges, and protruding nails that may pierce the tissue and damage the produce (Figure 2.3).

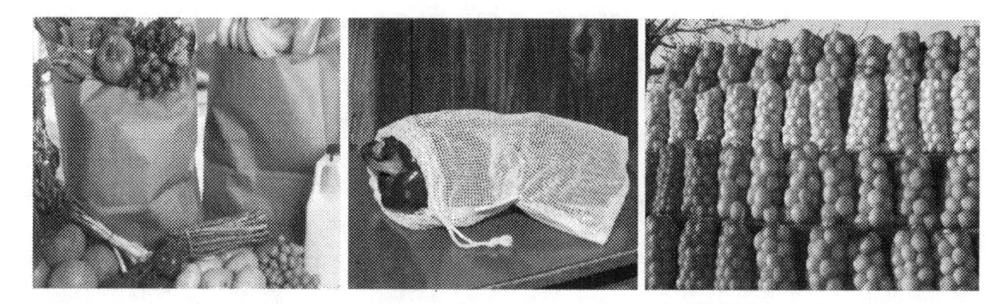

FIGURE 2.2 Various types of paper and mesh bags.

FIGURE 2.3 Wood and wood-based packaging materials.

2.2.3 CORRUGATED FIBERBOARD (CFB)

Corrugated fiberboards (CFBs), of varying capacities, are made by combining various layers of fiberboards together. These are most widely, commercially acceptable forms of bulk packaging for respiring fresh produce such as capsicum, bell peppers, etc., because of their relativity low price and versatility. They are reusable, recyclable, and have exceptional printability characteristics. In addition, these have a non-abrasive surface and provide excellent cushioning properties. However, these may get deformed under the influence of extreme pressure during stacking. Moreover, these are not resistant to rain and high humidity conditions as water may seep in and damage the products inside the box. Newer moisture resistant wax and plastic-based coatings have now been used to reduce moisture penetration to the package (Figure 2.4).

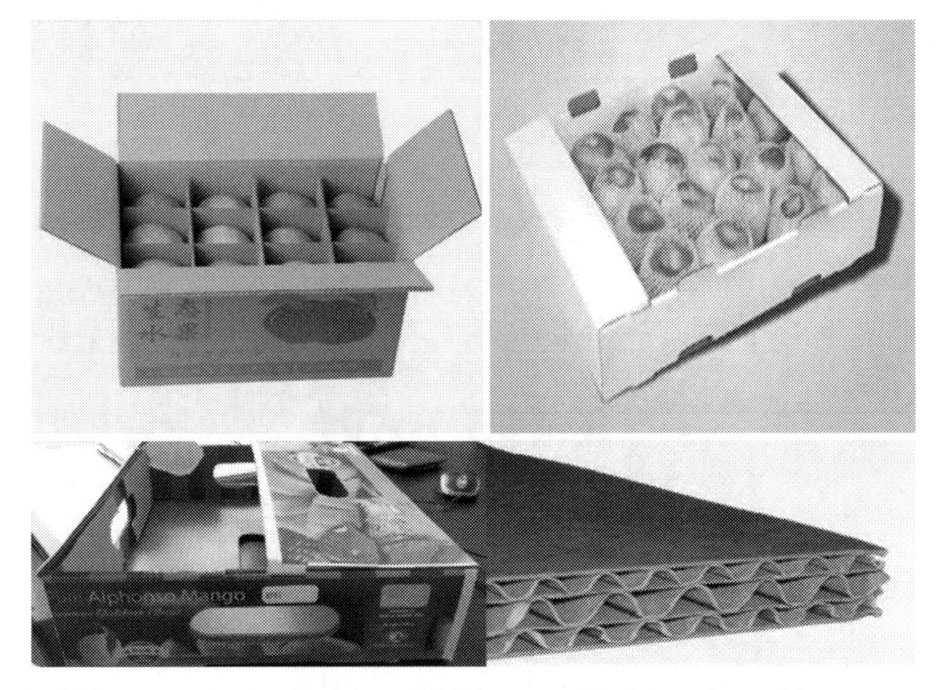

FIGURE 2.4 Various configurations of CFB boxes used for fruits and vegetables.

2.2.4 PLASTICS

Plastic (particularly, polyethylene) are undoubtedly the most widely and commercially accepted material for retail packaging of fresh fruit and vegetables. Their advantages in terms of strength and barrier properties include their transparency, durability, easy machinability, relatively lower cost, gas permeation properties, resistance to moisture, etc. Since each produce item has its own unique requirement for environmental gases, modified atmosphere (MA) packaging material must be specially engineered for each item (Kumar et al., 2015). Sufficient literature is available which points to the fact that plastics are the most desirable materials for short-term storage of fresh produce. These include rigid plastics such as clamshells, which are inexpensive, provide excellent protection to the produce, and a versatile consumer package for berries, mushrooms, etc., or products that are easily susceptible to damage by crushing. Molded polystyrene pallet bins and corrugated polystyrene containers have now been used as a replacement to wooden pallet bins. These are easy to clean, recycle, water-resistant, have a longer life, and can be made collapsible. Another form of plastic packaging for packaging of fresh horticultural produce include the flexible films such as Polyethylene (LDPE), polyvinyl chloride (PVC), polypropylene (PP) and cellulose acetate (CA) films. These films are semi-permeable, odorless, and allow the exchange of gases for respiration of the product. Moreover, trays made of molded pulp tray with individual cavities are used to avoid abrasion and bruising during transportation and storage since they ensure cushioning effect to the product. Plastic Punnets are recyclable, strong, and clear stackable containers, made of PET, PVC, or PP, that allow product visibility and are provided with perforations for ventilation, thus keeping the produce fresh. Extruded and Woven stretchable plastic net bags are generally made up of high-density polyethylene (HDPE) or Polyamide (PA; Nylon) that allow air circulation and easy diffusion of gases in and around the produce. They also reduce pack condensation, thus preventing microbial spoilage.

Plastic crates (made up of HPDE or PP) are another form of bulk packaging and transportation that are easily recycled and may be reused several times. These are easily cleanable, have high compressibility, stacking strength, and also allow adequate respiration through holes present in them to keep the produce fresh. Another form of packaging is shrink-wrapping of F&V in which individual unit is packed in a clear stretch film to reduce shrinkage and mechanical damage. A good film must be stretchable, elastic, and must strongly cling to the product. However, such wraps restrict adequate ventilation and respiration. Shrink wrapping has been used productively used for short term packing of cucumbers, potatoes, apples, onions, etc. (Figure 2.5).

FIGURE 2.5 Forms of plastic packages available in the market.

2.3 ADVANCED PACKAGING TECHNOLOGIES

Modern techniques of packaging of fresh or minimally processed F&V have opened opportunities for extending the shelf life long enough to preserve the produce and maintain its esthetic, sensory as well as nutritional properties. These tend to behave as smart and active system that incorporates both newer and conventional materials, with superior properties thus adding value and credibility to the packaging supply chain. The examples include modified atmosphere packaging (MAP), active packaging, and intelligent/smart packaging.

2.3.1 MODIFIED ATMOSPHERIC PACKAGING (MAP) OF FRUITS AND VEGETABLES (F&V)

This is an advanced packaging technique that consists of enclosing fresh horticultural produce in polymeric materials, in which the gaseous environment is actively or passively modified to reduce moisture evaporation, slow down respiration and senescence, and/or extend the shelf life of the stored product. F&V undergo respiration even after they are detached from their parent plant; as a result, the gases need to be diffused through the packaging films. Normal atmosphere comprises of 78% nitrogen, 21% oxygen, 0.03% carbon dioxide, trace elements, and water vapor. Divergence from this equilibrium is termed as MA. This modification can reduce respiration, decrease ethylene production, and action, retard tissue ripening, and softening, retard chlorophyll degradation and biosynthesis of carotenoids and anthocyanins, reduce enzymatic browning, and chilling injury, retard development of decay, maintain nutritional quality of produce and increases the overall acceptability of the produce (Aharoni et al., 2007). Each gas plays an important role while storage or packaging. Carbon dioxide slows down growth of most aerobic organisms, by reacting with free water inside the produce to form milk carbonic acid and thus reducing the pH of the product. This lowering down of pH negatively affects the metabolic activity of the produce as well as the related organisms. Bacterial and fungal growth is generally observed to be inhibited at a minimum 20% of carbon dioxide. Nitrogen, being inert, is used as a filler gas to prevent the collapse of packages containing high-moisture foods. Likewise, the presence of oxygen, inadequate quantity, is essential to maintain respiration. Increase in carbon dioxide and significant reduction in proportion of oxygen within specific limits preserves the original quality and extends storage life of F&V. Thus, the materials must be selectively permeable to allow exchange of gases during respiration (Figure 2.6).

Polyvinyl chloride (PVC), polyethylene terephthalate (PET), polyethylene (PE), low-density polyethylene (LDPE), linear low-density polyethylene (LLDPE), HDPE, PP, polyvinylidene chloride (PVDC), polyamide (PA) and polystyrene (PS) are the most commonly used polymer-based flexible films for MAP of F&V (Mangaraj et al., 2009). They may also be utilized as a component in rigid or semi-rigid packages, for example, as a liner inside a carton or as lidding on a cup or tray. However, the choice of material largely depends on the respiration rate of the produce as well as its desired storage temperature and humidity. PE and PP (BOPP, in particular), well suited for recycling and reuse, possess a significant combination of properties, such as flexibility, durability, resistance to water and moisture, and easy machinability. PVC based films are formed by addition polymerization of vinyl chloride, with several plasticizers to produce a heat sealable, resilient, and a clear film with high gas transmission rate. PET is explicitly and popularly used as packaging film with adequate gas barrier properties for modified packaged respiring products. PA-based films have excellent mechanical and thermal properties similar to PET and have similar application. However, owing to its relatively high cost, it is often co-extruded with other polymers. PS forms brittle, thermoplastic films and shows a poor barrier towards water vapor, gases, and aroma. They are widely used as transparent thermoformed trays. Multilayer films, often combined through lamination, metallization, or co-extrusion, seem to provide exceptional physical, chemical, as well as mechanical barrier in the modified atmosphere package. Metalized films have significantly enhanced barrier characteristics, and are usually chosen for this reason (Siracusa et al., 2017). Coating is another method to improve the functionality of the plastic films by imparting moisture resistance and heat sealability and improving barrier properties (e.g., PVDC copolymer coatings). Current commercial applications of MAP include packaging of apples, kiwis, pears, cherries, cabbage, bottle guard, broccoli, bananas, strawberries, etc., under various combinations of atmospheric storage gases.

Figure 2.7 summarizes the various criteria on which the selection of packaging material to be utilized for respiring produce is based.

FIGURE 2.6 Modified atmosphere packages available in market.

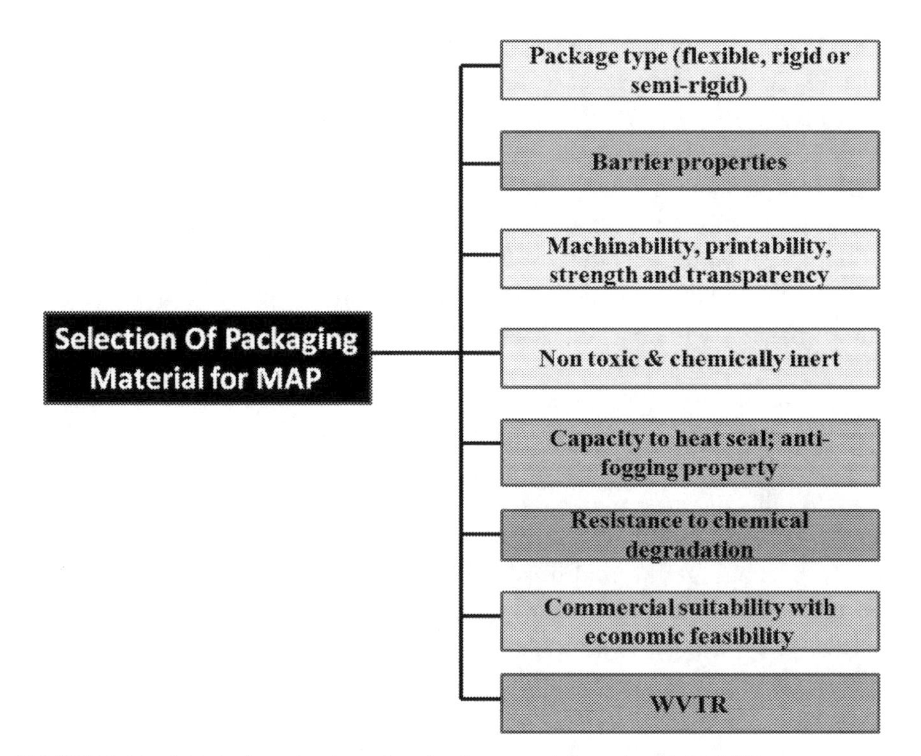

FIGURE 2.7 Criteria for selection of packaging material employed for MAP.

In addition to petroleum-based materials, adequate emphasis is also given to biodegradable polymers that are derived from starch and cellulose-based renewable sources such as agricultural residues, animal, and microbial sources and marine-based industrial wastes. Examples include polylactides (PLA), polyhydroxyalkanoate (PHA), polyhydroxy butyrate (PHB), etc. Moreover, chitin based biodegradable films, derived from the chitin of crustacean exoskeleton; also serve well as biodegradable edible coating material on the produce. This facilitates inhibition of moisture and gas migration, and improvement in the mechanical integrity and handling characteristics of the food product, which aims to accomplish MAP conditions.

2.3.2 ACTIVE AND INTELLIGENT PACKAGING OF FRUITS AND VEGETABLES (F&V)

Active packaging is an emerging packaging technique that is defined as the inclusion of an active system into packaging film or a container to maintain the quality or extend the shelf life of the product; these active systems include moisture controllers, oxygen scavengers, carbon dioxide emitters/absorbers, antioxidants, ethylene absorbers, flavor releasing/absorbing systems, antimicrobials, etc., (Bhardwaj et al., 2019). The terms active packaging and intelligent packaging are somewhat closely related. Intelligent or smart packaging usually involves the ability to sense or measure a property of the product, the inner atmosphere of the package and the shipping environment (Das, 2015), which can communicate the consumers about its condition and trigger active functions accordingly. Active packaging employs a packaging material that interacts with the internal atmosphere to prolong the shelf life of the produce. In the case of fresh horticultural produce, Ethylene scavengers, Oxygen scavengers, Carbon dioxide releasers, and antimicrobial agents are the most commonly used active systems. Accumulation of ethylene causes the yellowing of green vegetables as well as several postharvest diseases in fresh produce. Ethylenes scavengers are chemical reagents, incorporated into the packaging film and trap ethylene produced by fruit or vegetables. The presence of excess of oxygen triggers abnormal color and flavor changes, loss of nutritional properties and also, microbial spoilage. Oxygen scavengers may be employed either alone or in combination with MAP. For an oxygen absorber to be effective, the packaging film needs to have an oxygen barrier of intermediate performance (20 ml/m.d.atm), otherwise the scavenger will quickly become saturated and lose its ability to trap oxygen (Cruz et al., 2017). High carbon dioxide levels are desirable in some food packages such as strawberries, raspberries, etc. Since carbon dioxide is more permeable through plastic films as compared to oxygen, active production of carbon dioxide is needed to maintain the desired atmosphere in the package. Edible coatings and films also provide a vast array of advantages like biodegradability, edibility, good esthetic appearance as well as barrier properties, involving incorporating of bactericidal agents or growth inhibitors into film or coating-forming materials to form antimicrobial films and coatings (Bhardwaj et al., 2019).

Plastic packaging materials such as LDPE, PE, PVC, when impregnated with potassium permanganate, silica gel, alumina, natural zeolites, vermiculite, activated carbon acts as ethylene scavenging films for enhancing the shelf life of several fruits such as banana, avocado, honeydew melon and pear, banana, strawberry, lettuce, Chinese cabbage, kiwifruit, custard apple, mango, etc., to an appreciable extent (Álvarez-Hernández et al., 2018). Similar studies have been conducted by impregnating iron-based moisture absorbers as well as oxygen absorbers in films for guava, apples, avocadoes, peaches, tomatoes, thereby keeping them fresh (Murmu, 2018; Day, 1993).

Intelligent packaging enhances the functionality of the produce, by indicating changes in the package under the influence of heat, humidity, time, vibration, acidity, shock, light, microorganisms. Lettuce and pre-cut salads undergo rapid deterioration due to their high rate of respiration. Commercially novel breathable polymer packaging films are used for fresh-cut products. Intelligent packaging can change color or other properties and indicate about the freshness of the produce. Time-temperature indicators (TTI's) are systems that exhibit an irreversible change in a physical property (such as usually color, shape) under the influence of temperature change (Biegańska, 2017; Mijanur et al., 2018). However, ardent investigations still need to be carried to study their release profiles under different environmental conditions so that their efficiency is enhanced even further.

2.4 BIO-BASED AND EDIBLE PACKAGING: NEED OF THE TIME

Increasing demand for novel, eco-sustainable materials such as bioplastics and edible packaging for fresh horticultural produce has indeed revolutionized the world of food packaging. These materials can be developed with variable gas and moisture permeability, thus making them suitable for F&V. In addition to starch and cellulose-based materials, bioplastics based on poly-lactic acid (PLA), poly-hydroxy-alkanoates (PHA's), and poly-hydroxy-butyrates (PHB's) have also demonstrated exceptional biodegradable as well as barrier properties (Mistriotis et al., 2016; Ballesteros et al., 2018). However, these materials should be able to preserve and extend the shelf life of respiring produce long enough to justify their high cost as compared to cheaper petroleum-based alternatives. Edible coatings provide an additional protective barrier to produce and aid in modifying internal gas composition (Dhall, 2013). Proteins (e.g., soy protein isolate, whey protein isolate), lipids (neutral lipids, fatty acids, resins, etc.), and polysaccharides (starch, cellulose, chitosan, pectin, alginate, etc.), based edible coatings have achieved considerable attention with several examples of applications in F&V (Rojas-Grau et al., 2007). Another milestone is the successful incorporation of active and antimicrobial ingredients such as essential oils, extracts, etc., that can be incorporated into the polymer matrix and consumed along with food, hence improving nutritional and sensory properties (Ding and Lee, 2019; Ju et al., 2019). Food grade wax-based edible coatings (e.g., beeswax, candelilla wax), incorporated with functional ingredients (e.g., antimicrobial agents, antioxidants, etc.), for fresh produce, also present a novel method to preserve the produce (Ahmed et al., 2019). However, the compliance of these bio-based materials with quality and safety as food contact materials still remains a major concern for the manufacturers.

KEYWORDS

- **cellulose acetate**
- **corrugated fiberboard**
- **expanded polystyrene**
- **high-density polyethylene**
- **low-density polyethylene**
- **modified atmosphere packaging**
- **polyvinylchloride**

REFERENCES

Aharoni, N., Rodov, V., Fallik, E., Afek, U., Chalupowicz, D., Aharon, Z., Maurer, D., & Orenstein, J., (2014). Modified atmosphere packaging for vegetable crops using high water vapor permeable films. In: Wilson, C. L., (eds.), *Intelligent and Active Packaging for Fruits and Vegetables*. CRC Press, Taylor & Francis Group.

Ahmed, A., Ali, S. W., Imran, A., Afzaal, M., Arshad, M. S., Nadeem, M., Mubeen, Z., & Ikram, A., (2019). Formulation of date pit oil-based edible wax coating for extending the storage stability of guava fruit. *Journal of Food Processing and Preservation.* https://doi.org/10.1111/jfpp.14336.

Álvarez-Hernández, M. H., Artés-Hernández, F., Ávalos-Belmontes, F., Castillo-Campohermoso, M. A., Contreras-Esquivel, J. C., Ventura-Sobrevilla, J. M., & Martínez-Hernández, G. B., (2018). Current scenario of adsorbent materials used in ethylene scavenging systems to extend fruit and vegetable postharvest life. *Food and Bioprocess Technology, 11*(3), 511–525.

Ballesteros, L. F., Michelin, M., Vicente, A. A., Teixeira, J. A., & Cerqueira, M. Â., (2018). Use of lignocellulosic materials in bio-based packaging. In: *Lignocellulosic Materials and Their Use in Bio-Based Packaging* (pp. 65–85). Springer, Cham.

Bhardwaj, A., Alam, T., & Talwar, N., (2019). Recent advances in active packaging of agri-food products: A review. *Journal of Postharvest Technology, 7*(1), 33–62.

Biegańska, M., (2017). Shelf life monitoring of food using time-temperature indicators (TTI) for application in intelligent packaging. *Towaroznawcze Problemy Jakości (Polish Journal of Commodity Science), 2*, 75–85.

Cruz, R. S., Soares, N. D. F. F., & Andrade, N. J. D., (2007). Efficiency of oxygen: Absorbing sachets in different relative humidities and temperatures. *Ciência e Agrotecnologia (Science and Agrotechnology), 31*(6), 1800–1804.

Das, M., (2015). Nanocomposites in food packaging. In: Mohanty, S., Nayak, S. K., Kaith, B. S., & Kalia, S., (eds.), *Polymer Nanocomposites based on Inorganic and Organic Nanomaterials.* Scrivener Publishing, Wiley.

Day, B. P. F., (1993). Fruit and vegetables. In: *Principles and Applications of Modified Atmosphere Packaging of Foods* (pp. 114–133). Springer, Boston, MA.

Dhall, R. K., (2013). Advances in edible coatings for fresh fruits and vegetables: A review. *Critical Reviews in Food Science and Nutrition, 53*(5), 435–450.

Ding, P., & Lee, Y. L., (2019). Use of essential oils for prolonging postharvest life of fresh fruits and vegetables. *International Food Research Journal, 26*(2).

Ju, J., Xie, Y., Guo, Y., Cheng, Y., Qian, H., & Yao, W., (2019). Application of edible coating with essential oil in food preservation. *Critical Reviews in Food Science and Nutrition, 59*(15), 2467–2480.

Kumar, V., Shankar, R., & Kumar, G., (2015). Strategies used for reducing postharvest losses in fruits and vegetables. *Int. J. Sci. Eng. Res., 6*, 130–137.

Mandal, G., (2015). Value addition of fruits and vegetables through packaging. In: Sharangi, A. B. et al., (eds.), *Value Addition of Horticultural Crops: Recent Trends and Future Directions* (pp. 191–199).

Mangaraj, S., Goswami, T. K., & Mahajan, P. V., (2009). Applications of plastic films for modified atmosphere packaging of fruits and vegetables: A review. *Food Engineering Reviews, 1*(2), 133.

Mijanur, R. A., Kim, D., Jang, H., Yang, J., & Lee, S., (2018). Preliminary study on biosensor-type time-temperature integrator for intelligent food packaging. *Sensors, 18*(6), 1949.

Mistriotis, A., Briassoulis, D., Giannoulis, A., & D'Aquino, S., (2016). Design of biodegradable bio-based equilibrium modified atmosphere packaging (EMAP) for fresh fruits and vegetables by using micro-perforated poly-lactic acid (PLA) films. *Postharvest Biology and Technology, 111*, 380–389.

Murmu, S. B., & Mishra, H. N., (2018). Selection of the best active modified atmosphere packaging with ethylene and moisture scavengers to maintain quality of guava during low-temperature storage. *Food Chemistry, 253*, 55–62.

Rojas-Grau, M. A., Grasa-Guillem, R., & Martin-Belloso, O., (2007). Quality changes in fresh-cut Fuji apple as affected by ripeness stage, anti-browning agents, and storage atmosphere. *Journal of Food Science, 72*, 36–43.

Siracusa, V., Dalla, R. M., & Iordanskii, A., (2017). Performance of poly(lactic acid) surface modified films for food packaging application. *Materials, 10*(8), 850.

Thompson, A. K., (1996). *Postharvest Technology of Fruit and Vegetables*. Blackwell, Oxford.

CHAPTER 3

PACKAGING AND STORAGE OF FRUITS AND VEGETABLES FOR QUALITY PRESERVATION

KUNDAN KISHORE and DEEPA SAMANT

Central Horticultural Experiment Station (ICAR-IIHR), Aiginia, Bhubaneswar–751019, Odisha, India

ABSTRACT

Packaging and storage has become an integral part of the horticulture production system due to its significance in improving shelf life and minimizing postharvest loss. The future of packaging is one in which the package will increasingly operate as a smart system incorporating both smart and conventional materials, adding value and benefits across the packaging supply chain. Packaging should be low cost and environmentally safe. In order to ensure better marketability higher package quantity is needed for fresh F&V active packaging and intelligent packaging are emerging and exciting areas of food preservation technique, which can confer many preservation benefits on a wide range of foods. The aim of active packaging is to match the properties of the package to the more critical requirements of the food. Adoption of some of these methods will require changes in attitude to packaging and a willingness to address regulatory issues where chemical effects are used. Application of new emerging technologies offers greater potential in horticulture sector. The intelligent system is an effective approach not only to maintain the quality or value of a product but also to provide information about produce quality to consumers.

3.1 INTRODUCTION

Fruits and vegetables (F&V) are not only the most important components of the horticulture industry of the world but also essentially important in the human diet as they are rich sources of proteins, carbohydrates, vitamins, dietary fibers, minerals, and antioxidants. The fresh fruit industry provides great opportunities to boost the economy of the country through export diversification and revenue generation. In developing countries like India, inadequate infrastructure and poor postharvest management practices are the major challenges for the horticultural industry as they are responsible for quantitative and qualitative losses of fruits. In addition, both the factors significantly affect market potential, export potential and farm income (Wu, 2010). It is estimated that every year roughly one-third of the food produced for human consumption worldwide is lost or wasted (FAO, 2011). In India, sizable postharvest loss (>30%) is a matter of great concern. It is estimated that more than 70 million tons of vegetables and fruits are lost annually in India, which accounts for the monetary loss of $33,745 (ASSOCHAM, 2013). The loss is so huge that it surpasses the annual budget allocation for the agriculture sector. In most of the tropical F&V rapid rate of biochemical reactions occurs during ripening process which is considered as the most important limiting factors in the marketing of fresh F&V. The harvested produce should be instantly subjected to postharvest management practices to minimize economic loss. The application of appropriate postharvest technologies is a pragmatic approach to reduce postharvest losses and maintain the quality of fresh tropical fruits. Postharvest quality of horticultural produce is affected by various factors such as harvesting at suboptimal stage, improper packaging, and physical damage during transportation, high temperature, high relative humidity (RH), and microbial activities. Packaging and storage management are key factors to ensure quality and better shelf life of produce. Among these factors, packaging, and storage are crucial components of postharvest management as they take into account respiration rate, ethylene production, temperature, humidity, and concentration of gases in order to enhance shelf life of F&V (Kader, 2005). Postharvest losses (PHT) not only affect the economic viability of horticulture production system but also limit the availability of produce. In order to minimize the PHL and enhancement of shelf life, various packaging, and storage technology is used.

3.2 FACTORS AFFECTING SHELF LIFE

High perishability is an inherent behavior of F&V which may be attributed to their natural properties, mechanical damage, external factors, microbial contamination, and the level of postharvest management. Although it is difficult to control all the factors responsible for affecting shelf life, the intensity of influence of these factors may be minimized.

3.2.1 INHERENT PROPERTIES

F&V are still functionally alive even after harvesting and continue their physiological activities like respiration, transpiration, ethylene production and other cellular changes that cause loss of texture, fruit color, flavor, and nutritional value (Atanda et al., 2011). Postharvest losses of tropical F&V are mainly due to their following inherent properties (Kusumaningrum et. al., 2015):

- High moisture content (> 80%) causes high transpiration loss;
- Large fruit size (jackfruit, pumpkin, durian, pineapple) responsible for more damage during transportation;
- High respiration rates lead to heat emission at ambient temperature;
- Soft skin vulnerable to mechanical injury.

3.2.2 EXTERNAL FACTORS

External conditions play a crucial role in influencing the quality and shelf life of fruits. Factors like temperature, RH, and composition of the air surrounding the produce substantially affect physiological activities inside the fruit. Generally, high temperature and low RH increases transpiration rate and respiration activity, which in turn affect the shelf life of fruits (Rene, 2001). Composition of CO_2, O_2 and ethylene plays vital role in influencing respiration rate and general metabolism of fruits. In order to reduce the rate of respiration, the concentration of O_2 should be low.

3.2.3 MICROBIAL ACTIVITY

Biological spoilage is one of the most important causes of postharvest loss. For growth of microorganisms, nutrients and water content are prerequisite. Since fresh F&V are rich in nutrients and water content, they are highly susceptible to microbial growth. In tropical regions, microbial spoilage is high due to the prevalence of optimized temperature for microbial growth. Approximately 20% of produce are affected by microbial spoilage during postharvest handling in tropical countries (Spadaro et al., 2002; Zhu, 2006).

3.2.4 HANDLING OF PRODUCE

Handling of fruit plays a vital role in ensuring fruit quality as it matters from harvesting to marketing. If fruits are not harvested at optimal maturity, it is likely that postharvest loss will be substantial. Handling methods determine the extent of physical injuries and consequently quality and marketing of fruit (Kader, 2002). Mechanical damage during handling not only affects appearance, fruit quality, and nutritive value but also renders fruits susceptible to microbial infection (Wu, 2010).

3.3 TECHNOLOGIES FOR SHELF LIFE EXTENSION

Various methods are used to extend the shelf life of fruits by maintain fruit quality within an acceptable range. Among these methods, packaging, and storage have a special place in postharvest management operation.

3.3.1 PACKAGING

Packaging plays a crucial role in the supply chain of horticultural produce. The key function of packaging is to facilitate efficient transport within the whole supply chain and ensure the quality of produce by preventing physical damage and by protecting against manipulation. Packaging also addresses the key issue of quality and safety of produce from production to final consumption by preventing unwanted chemical and biological changes. Thus packaging acts as a barrier to protect the item from environmental factors such as oxygen, moisture, pests, and diseases and any type of contamination like chemical and microbial (Yildirim, 2011; Arvanitoyannis and Oikonomou, 2012). As F&V possess high moisture content which cause their rapid deterioration in fruit quality under normal climatic condition. Moreover, fresh produce undergoes various physical and mechanical injuries during its long and complicated journey that starts from the grower's field and ends at the consumer. Any kind of injury during transit leads to spoilage and deterioration in quality of F&V, which in turn affect their quality, shelf life, and marketability. In addition to protection against spoilage, packaging provides containment of produce in convenient units for easy handling during their journey from field to table. F&V are transported to different places in various types of packaging materials. Wooden pallet boxes and fireboard cartons are the most common form of packaging of bulky horticultural produce like F&V. However, additional internal packaging with paper wraps or pads is required to reduce damage during transportation and handling. For delicate fruits, packaging materials have been suitably modified to reduce the intensity of damage. With the technological advancement in packaging sector (passive, active, and smart packaging), postharvest handling of horticulture produce has become both market-oriented and consumer-oriented (Figure 3.1). F&V are being packaged in consumer-friendly packs using various types of packaging material, viz., recycled paper pulp bag, mesh bag, plastic bag, shrink-wrap and rigid plastic packages, etc. The role of internal packaging is primarily passive in nature as it acts as a barrier between the surrounding environment of produce and the external environment. However, such packaging systems are limited in their ability to further extend the shelf life of the packaged item (Matche, 2013).

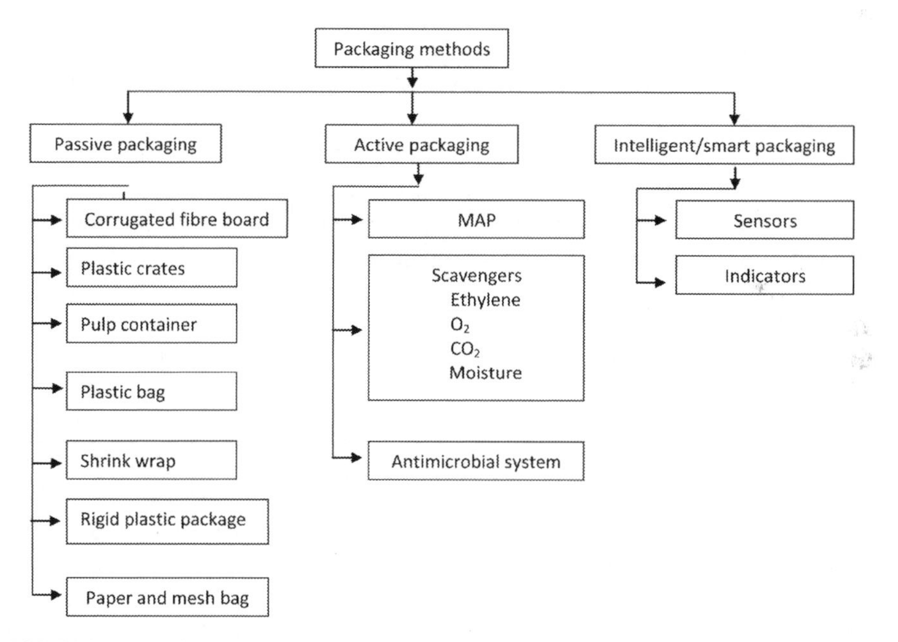

FIGURE 3.1 Packaging methods for fruits and vegetables.

3.3.1.1 PASSIVE PACKAGING

Wooden pallet boxes, corrugated fiberboard (CFB), plastic crates, and sacks are conventionally used for transportation of F&V. Wooden pallet boxes are cheap and can be reused. Moreover, standard-size pallets can effectively accommodate more produce in transportation. The use of wooden pallet boxes are gradually being replaced by CFB because of its relativity low cost and versatility. Moreover, it also has better strength and serviceability. In conventional passive packaging, cushioning materials are essentially required at different stages to prevent F&V from damage. Many types of cushioning materials such as newspaper, newspaper cuttings, rice straw, bubble sheet, foam nets, molded trays, gunny bags, etc., are being commonly used during packaging. Jarimopas et al. (2017) reported that bruising damage is one of the important factors for postharvest losses in apple throughout the supply chain. Individual wrapping of fruits with foam net and corrugated board effectively minimized the damage.

3.3.1.1.1 Pulp Container

Packaging made of synthetic polymers is being replaced by environment friendly biodegradable materials such as sugarcane bagasse and binders, paper pulp and starch binder (Sridach, 2014). These materials are mainly used for packaging of high value F&V. Pulp containers are not only available in a variety of shapes and sizes but also relatively inexpensive. They can absorb moisture from the surface of produce, and thus reduces the chance of spoilage due to high humidity and microbial growth.

3.3.1.1.2 Plastic Bags

The appearance, fruit texture, color, shelf life, disease, and disorder, and marketability of produce are remarkably influenced by packaging materials. Plastic bags (polyethylene film) are the most common packing material for horticultural produce as they are cost-effective and also facilitate easy examination of items. Plastic films are available in a wide range of thicknesses and grades, which are used to control the combination of gases like oxygen and carbon dioxide and water vapor inside the package. Polyethylene film of 20 μ thickness with holes effectively improves shelf life of mandarin by minimizing weight loss, retaining texture, and lowering pathological problems and in turn improves marketability (Bhattarai and Shah, 2017).

3.3.1.1.3 Shrink Wrap

One of the newest trends in produce packaging is the shrink-wrapping of individual produce items. Shrinkwrapping has been used successfully to package high value F&V. Shrink wrapping with an engineered plastic wrap reduces shrinkage, mechanical damage and protect the produce from diseases. The shelf life of cucumber can be increased for two weeks if individual frits are subjected to shrink-wrap and stored at 12 ± 1°C with 90–95% RH. Under such condition freshness, color, and texture of cucumber are retained (Dhall et al., 2012). Similar result was also obtained in apple when individual fruits were shrink-wrapped with 25-micron shrinkable film. The polymeric shrink-wrapping creates a safe protective coating over the fruit surface that reduces respiration rate and in turn, increases the shelf life of apples by at least two weeks (Thakur et al., 2017).

3.3.1.1.4 Rigid Plastic Packages

Rigid or semi-rigid plastic packages, also called clamshell, are mainly suitable for soft fruits. It is gaining popularity because of its low cost, versatility, and attractive appearance. Clamshells are most often used for high value items like cherry, grapes, strawberry, plum, mushrooms, etc., or items that are easily damaged by crushing during handling and transport. Such fruits are highly perishable due to high moisture content, soft skin and soft fruit texture. Moreover, they are highly prone to anaerobic spoilage. Hence they are packaged in clamshell with cover of cellophane, cellulose acetate (CA), polystyrene or other suitable film cover. Now a day recyclable clamshell is used for pre-cut fruit and vegetables, including salads.

3.3.1.1.5 Paper and Mesh Bags

Both types are bags are commonly used for consumer-friendly packing of potatoes, onions, tomato, garlic, sweet corn, chili, etc. Hoverer mesh bag has wider applicability as it facilitates air circulation, which extends shelf life of tuber and rhizomatous crops.

3.3.1.2 ACTIVE PACKAGING

Traditional food packaging is meant for protection, communication, convenience, and containment (Robertson, 2006). The package is used to protect the product from the deteriorative effects of external environmental conditionals like temperature, light, moisture, microbes, etc. Additionally, it also provides the consumer with the greater ease of use (Yam et al., 2005; Marsh and Bugusu, 2007). The key safety objective for traditional packaging materials is to keep the quality of produce intact and protect it from external factors. While the smart packaging systems like active and intelligent packaging concepts are based on the useful interaction between packaging environment and produce to provide ensure active protection as the shelf life of food items is influenced by biological, chemical, and physical interactions between the food, package, and the environment (Manesh and Azizi, 2017). Active packaging is an innovative approach to ensure the quality of produce as well as extended shelf life. In this packaging system material interacts with the environment in such a way that it either releases or absorbs substances into or from the packaged food or the environment surrounding the food to minimize metabolic activities and extend shelf life (EFSA, 2009). Modified atmospheric packaging and the use of scavenging materials are the most important component of active packaging.

3.3.1.2.1 Modified/Controlled Atmospheric Packaging (MAP)

Fresh F&V are unique food products that perform various physiological and biochemical processes, viz., ripening, respiration, transpiration, and senescence even after harvesting. On account of which, they could not maintain their freshness for a long period of time and suffer from heavy postharvest losses. These physiological and biochemical processes cannot be stopped, however, they can be slowed down by modifying the gas atmosphere surrounding the produce with the use of suitable packaging. Modified atmosphere packaging (MAP) is the way to modify the natural ambient air in the package by replacing a gas or gas mixture, often nitrogen and carbon dioxide. This packaging under a protective atmosphere preserves the quality of fresh produce over a longer period of time, and facilitates producers to access distant market. Packing of F&V with plastic films of different kinds of combinations of materials, perforation, and inclusions of chemicals is referred to as either MAP or controlled atmosphere packaging (CAP) depending on the precision in maintaining levels of O_2, CO_2, and other gases. CAP provides better precision than MAP; however, both types of packaging system follow the same principle. MAP and CAP are effective technology for packaging and storage of F&V (Kader, 1989).

The gaseous combinations around the produce get altered inside the polymeric films due to interaction between packaging and biological activities of fresh produce, namely respiration and transpiration. Depending upon the permeability of packaging and nature of horticultural commodity, the atmosphere around the perishables becomes rich in CO_2, whereas, the concentration of O_2 starts to fall. With the fall of O_2 levels below 10–12%, there starts decrease in respiration rate, which keeps on falling till the O_2 level becomes 2–5%. On the other hand, the elevated level of CO_2 above atmospheric level not only reduces respiration rate but also suppresses ethylene production, activities of microorganisms. However, temperature and RH influence the efficacy of MAP. Increasing the concentration of nitrogen (N_2) gas to 100% in MAP enhances shelf life of leafy vegetables like lettuce and cabbage. It has been observed that the gaseous combinations change with time even 100% N is used. The concentration of O_2 and CO_2 increases to 1.2 to 5.0% and 0.5 to 3.5%, respectively, after 5 days of storage (Kosaki and Itoh, 2002). The composition of air in MAP varies with the crop species and their nature of repining (Table 3.1). Thus packaging helps the fresh produce to retain its postharvest quality and marketability for a longer period of time. MAP is not only used in packaging, but they can feature as part of the production process. The standards required by MAP are comparatively high, and have to be controlled and monitored to ensure safety. Therefore, food manufacturers rely on modern MAP gas technology and various levels of quality assurance for maximum process safety.

TABLE 3.1 Optimized Composition of Gases for Modified Atmospheric Packaging of Fruits and Vegetables

Crops	O_2	CO_2	N_2	Crops	O_2	CO_2	N_2
Apple	1–2	1–3	95–98	Broccoli	1–2	5–10	88–94
Kiwi fruit	1–2	3–5	80–92	Beans	2–3	5–10	87–93
Peach	1–2	3–5	93–96	Cabbage	2–3	3–6	81–95
Pear	1–2	0–1	96–98	Carrot	5	3–4	91–95
Banana	2–5	2–5	90–96	Chili	3	5	92
Grapes	2–5	1–3	92–97	Cauliflower	2–5	2–5	90–96
Orange	5–10	0–5	85–95	Sweet corn	2–4	10–20	76–88
Papaya	2–5	5–8	87–93	Cucumber	3–5	0	95–97
Pineapple	2–5	5–10	85–93	Lettuce	1–3	0	97–99
Strawberry	5–10	15–20	70–80	Mushroom	3–21	5–15	65–92

3.3.1.2.2 Scavengers

Scavengers are chemicals or packing materials which reduces the activities of ethylene, oxygen, carbon dioxide, and microbes inside the package to slow down the respiration rate, ripening process, and microbial spoilage. Scavengers must be safe, easily handled, compact in size, and must not produce toxic substances or offensive odors/gases.

Scavengers can be divided into active scavenging systems (absorbers) and active-releasing systems (emitters). The former remove undesired compounds from the food or its environment, for example, moisture, carbon dioxide, oxygen, ethylene, or odor, whereas the latter add compounds to the packaged food or into the headspace, such as antimicrobial compounds, carbon dioxide, antioxidants, flavor, etc.

1. **Ethylene Scavenger:** Fruit and vegetables are primarily consumed for their nutritional and nutraceutical value. The postharvest life of fresh produce are affected by various factors, and among them, ethylene plays a major role even at low concentrations. It acts as a ripening agent by accelerating the rate of respiration and degradation of chlorophyll. On the basis of respiration behavior and rate ethylene production during maturation and ripening process, fruits are classified into climacteric and non-climacteric (Cherian et al., 2014). In case of climacteric fruits (mango, apple, papaya, avocado, sapota, kiwifruit, banana, pear, blueberry, jackfruit) a peak in both respiration and ethylene production occurs during ripening, whereas non-climacteric fruit (citrus, pineapple, grapes, melon, peas, pepper, cucumber) do not exhibit that dramatic change in respiration, maintaining ethylene production at relatively low level (Paul et al., 2012). In climacteric fruit, a high rate of ethylene production accelerates fruit softening, color changes, texture softening, alteration in sugar content and volatile aromas synthesis, whereas in non-climacteric fruits, ethylene stimulates senescence which is often associated with chlorophyll degradation (Barry and Giovannoni, 2007). In both climacteric and non-climacteric fruit, ethylene can induce chilling injuries and physiological disorders (Wills, 2015). The respiration rate and ethylene production in F&V have been described in Tables 3.2 and 3.3 (Rizvi, 1981; Kader, 2002).

   Ethylene concentrations higher than 4 ppm at 20°C causes significant reduction in shelf life of peach, avocado, and tomato, whereas less than 1 ppm ethylene may affect the shelf life of banana, strawberry, lettuce, Chinese cabbage, kiwifruit, custard apple and mango (Warton et al., 2000; Wills et al., 2001). To prolong shelf life and maintain an acceptable quality, accumulation of ethylene in the packages should be discouraged by ethylene absorbers. Potassium permanganate ($KMnO_4$) acts as an ethylene scavenger which is placed inside the package to reduce the concentration of ethylene. In order to facilitate efficient scavenging of ethylene, adsorption of $KMnO_4$ onto a porous inert material with a high surface area such as silica gel, zeolite, vermiculite, alumina, and activated charcoal is suggested. Green Pack, a sachet of $KMnO_4$ embedded in silica is commonly used for enhancing the shelf life of F&V. However, it has been reported that $KMnO_4$-based C_2H_4 scrubbers supported onto alumina nanoparticles have higher C_2H_4 removal rate than scrubbers based on SiO_2 nanoparticles (Spricigo et al., 2017; Álvarez-Hernández, 2018). Another type of ethylene scavenging activity is based on adsorption and subsequent breakdown of ethylene. New palladium-based ethylene scavenging technology is effective in prolonging the shelf life of fresh fruits (Smith et al., 2009).

TABLE 3.2 Respiration Rate of Fruits and Vegetables

Respiration Rate ($mLCO_2$ $kg^{-1}h^{-1}$ at 5°C)	Crops
Very low <5	Nut, dried fruits and vegetables
Low 5–10	Apple, pear, kiwifruit, grape, potato onion, garlic
Moderate 20–40	Carrot, cabbage, cherry, lettuce, peach, plum, pepper, tomato, cauliflower
High 40–60	Apricot, avocado, papaya, custard apple, Brussels sprouts, green onion
Very high >60	Strawberry, blackberry, raspberry, asparagus, broccoli, mushroom, pea, spinach, sweet corn

Source: Kader and Saltveit (2005). Used with permission.

2. **Oxygen Scavenger:** F&V are highly sensitive to oxygen as it is directly or indirectly responsible for their deterioration. In fact, in many cases, deterioration is caused by oxidation reactions or by the presence of spoilage aerobic microorganisms. Therefore, in order to preserve these products, oxygen is often excluded. MAP is one of the viable tools to reduce O_2 content inside food packaging. However, for many food items, the levels of residual oxygen that achieved by

regular (MAP) technologies are too high for maintaining the desired quality and achieving the desired shelf life (Damaj et al., 2009). The oxygen scavenging packaging materials substantially reduces the oxygen inside the packaging material which is otherwise not achieved by MAP (Zerdin et al., 2003). An oxygen scavenger is a material or combination of reactive compounds incorporated into a package structure that may combine with oxygen and effectively remove the oxygen from the inner package environment. Iron powder (metallic), ascorbic acid, and catechol (non-metallic) are commonly used O_2 scavengers. They also include enzymatic scavenger systems using either glucose oxidase or ethanol oxidase (Day, 2003). Oxygen-scavenging compounds can be incorporated directly into the packaging material or can be placed in the form of the sachet. These materials include flexible films, rigid plastics and liners (Wong et al., 2017).

TABLE 3.3 Ethylene Production Rate of Fruits and Vegetables

Ethylene production rate (μL/ kg. hr at 20°C)	Crops
Very low (less than 0.1)	Citrus fruits, strawberry, artichoke, cauliflower, cherry, asparagus, potato, grape, leafy vegetables, root vegetables
Low (less than 1.0)	Pineapple, blueberry, raspberry, persimmon, cucumber, sweet pepper, chili, pepper, pumpkin, watermelon
Moderate (less than 10)	Banana, fig, mango, melon, mango, tomato
High (less than 100)	Ripe kiwifruit, apple, avocado, plum, peach, apricot, nectarine, papaya
Very high (More than 100)	Custard apple, passion fruit

3. **CO_2 Scavenger:** Carbon dioxide gas is useful for the modified atmosphere packaging of foods as it reduces the respiration rate its excess accumulation may be detrimental to the quality of the product. In order to prevent spoilage, the level of CO_2 should be maintained within the limit. Under such condition, the use of CO_2 scavengers is advisable for preserving the food quality and package integrity. The common mechanisms of CO_2 absorption in food packages are non-harmful chemical reactions (absorbent) and adsorption. Oxide and hydroxide of calcium and sodium carbonate absorb excess CO_2 by performing chemical reactions, whereas zeolite and activated

carbon are used as CO_2 adsorbent. These materials can be enclosed in a sachet and can conveniently be placed in the food package.

4. **Moisture Scavengers:** Moisture content and water activity are critical factors affecting the quality and shelf life of produce (Labuza and Hyman, 1998). For instance, many dry fruits, vegetables are sensitive to humidity during storage, and even low RH levels inside the packages may cause significant quality deterioration. Increase in moisture makes the products more prone to microbial spoilage and may cause alterations in texture and appearance, consequently reducing shelf life (Day, 2008). For fresh fruit/vegetables, keeping a controlled high RH level inside the package is beneficial in preventing drying. Moisture absorbent pads, sheets, and blankets are used for controlling excessive moisture content in F&V. Desiccant like silica gel are used to control humidity in the packaging headspace.
5. **Antimicrobial System:** It is a novel approach in active packaging system wherein controlled release of antimicrobials from packaging materials is done. Antimicrobials could extend the shelf life of produce by preventing bacterial growth and in turn, spoilage. Recent advancement has been made in this system wherein antimicrobial substance is released on command when bacterial growth occurs. This system is known as "BioSwitch" (De Jong et al., 2005). The efficacy of antimicrobial system is influenced by environmental factors like pH, temperature, and light intensity. These external factors act as stimulus for the release of the antimicrobial component in the package. In this system, the antimicrobial is released on command, and the system is active only at specific conditions. This system has not only the potential to increase the stability and specificity of preservation but also to reduce the chemical load on food items. A common example of release on command antimicrobials in food packaging is the inclusion of polysaccharide or protein particles that encapsulate antimicrobial compounds. With the increase in microbial population inside the package, polysaccharides will be utilized by bacteria, and finally, antimicrobial compounds will be released, which will prevent microbial growth.

3.3.1.3 INTELLIGENT PACKAGING SYSTEM

Awareness about the consumption of safe food has steered innovations in packaging technologies. Intelligent packaging system deals with the monitoring of quality aspects of produce and report information to the consumer. According to the European council, intelligent packaging monitors the condition of packaged food or the environment surrounding the food. Intelligent packaging is a novel tool to detect, sense, and record the changes occurs in the produce with time and environmental condition (Restuccia et al., 2010; Realini and Marcos, 2014). The purpose of the intelligent system is to maintain the quality or value of a product and to provide information about produce quality to consumers (Robertson, 2006). In contrary to scavengers, intelligent components do not release their constituents into the food but to detect the changes. This system can play a significant role in the improvement of hazard analysis and critical control points' (HACCP) and quality analysis and critical control points' (QACCP) systems, which are developed to onsite detection of unsafe food, identify potential health hazards and establish strategies to reduce or to eliminate their occurrence (Heising et al., 2014; Biji et al., 2015). Basically, there are three tools for intelligent systems; sensors, indicators, and radiofrequency identification (RFID) (Kerry et al., 2006; Vanderroost et al., 2014).

3.3.1.3.1 Sensors

Sensors detect and measure the activities of microbes and gases inside the package (Kerry et al., 2006). Various types of sensors such as gas sensor, biosensor, and chemical sensor are used in the processing industry. Gas sensors are used for the detection of gaseous combinations in the package. It includes oxygen sensors (detection of O_2 concentration), carbon dioxide sensors (detection of CO_2 concentration), water vapor sensor (detection of humidity level), ethanol sensor ((detection of anaerobic reaction), metal oxide semiconductor field-effect transistors, organic conducting polymers and piezoelectric crystal sensors, etc., (Kress-Rogers, 1998; Kerry et al., 2006). On the other hand, biosensors are mandated to detect, record, and transmit information pertaining to biological reactions (Yam et al., 2005). The bioreceptors may be either organic or biological materials like enzyme, hormone, nucleic acid, antigen, microbes, etc. Food Sectinel System® (SIRA Technologies Inc.) is a commercial biosensor developed to detect the food pathogens, whereas ToxinGuard® (Toxin Alert, Canada) is visual diagnostic system based on antibodies printed on polyethylene-based plastic packaging material which detect the targeted pathogens such as *Salmonella* sp., *Campylobacter* sp., *E coli.*, *Listeria* sp. (Bodenhamer et al., 2004). The chemical sensor or the receptor is a chemical selective coating capable of detecting the presence, activity, composition, concentration of particular chemical or gas through surface adsorption. Carbon nanomaterials like, graphene, graphite, nanofibers, and nanotubes are applied in chemical sensors because of their excellent electrical and mechanical properties along with the high specific surface area (Vanderroost et al., 2014). These nano-based sensors can be used to detect pathogens, chemical contaminants, spoilage, product tampering, track ingredients or products through the processing chain (Liu et al., 2007; Nachay, 2007; De-Azeredo, 2009).

3.3.1.3.2 Indicators

Indicators are substances that indicate the presence, absence, or concentration of undesirable substance or the degree of reaction between two or more substances by means of a characteristic change, especially in color (Hogan and Kerry, 2008). Freshness indicators provide the information on status of product resulting from chemical changes and microbial growth within food products. The reaction between the metabolites released by microbes and the integrated indicators within the package provide visual information regarding the microbial status of the product (Kerry et al., 2006; Kuswandi et al., 2013). A colorimetric chitosan bio-based pH indicator has been developed to detect microbial growth (Yoshida et al., 2014) On the other hand, time-temperature indicators (TTIs) provide information on the status of threshold temperature. It also indicates the minimum amount of time a product has spent above the threshold temperature. These indicators provide visual indications of temperature history during packing, storage, and distribution. Basically, TTIs are small tags or labels that keep track of time-temperature of a perishable commodity from the production point to distribution point. The *ripeSense*™ is the world's first intelligent ripeness indicator label which provide information on important aspects of fruit quality (Scetar et al., 2010). A number of smart packaging technologies have been developed in recent years, which are being integrated into the packaging systems not only to improve shelf-life but also to ensure quality of produce in the supply chain and keeping the consumer inform about the status of produce.

3.4 STORAGE

For maintaining quality and extending shelf life of F&V, harvesting at optimum maturity, minimization of mechanical injuries, precooling, packaging, and storage at optimum temperature are crucial factors (Kader et al., 1989). Proper storage of F&V plays a crucial role in the management of postharvest loss and in achieving higher returns, which in turn act as a driver for the development of the horticulture industry. Storage condition depends on the intrinsic physiological features of produce. There is a wide range of variation in the shelf life of produce which is mainly depends on water content, transpiration rate, respiration rate and ethylene production rate in the produce. Fruits like strawberry, jamun, raspberries, and grapes have short shelf life, whereas onions, potatoes, garlic, and pumpkins may be kept for weeks without being damaged. Storage conditions also depend on specific product characteristics. For example, temperate fruits and leafy vegetables tolerate temperatures close to 0C, whereas most tropical fruits exhibit chilling injury at low temperature. It gives an indication that temperate fruit are adapted to develop and ripe at low temperature.

3.4.1 LOW TEMPERATURE

When tropical F&V are harvested, they possess high internal temperature, which causes water loss and increase in metabolic activities. If harvested crops are not subjected to low temperature (pre-cooling), their quality and shelf life is substantially reduced. Precooling is a process that reduces field heat from freshly harvested produce like F&V. It can be done by room cooling, hydro-cooling, forced-air cooling, and vacuum cooling. Forced air-cooling system and hydro-cooling are usually practiced for export items (Wu, 2010; Chinaphuti, 2011). Imposition of low temperature to F&V reduces their respiration rate and in turn, metamorphism. This process helps in controlling the activity of microbes and spoilage of produce and thus enhances their shelf life. Optimum temperature range for F&V has been described in Table 3.4.

TABLE 3.4 Optimum Storage Condition for Fruits and Vegetables

Crops	Temperature Range (°C)	Relative Humidity	Storage Time
	Temperate Fruits		
Apple	–1.0–4.5	90–95	4–32 weeks
Apricot	–0.5–0.0	85–95	1–3 weeks
Blackberry	–0.5–0.0	85–100	2–3 days
Blueberry	–0.5–0.0	95–100	2 weeks
Cherry	–1.0–0.0	90–95	2–4 weeks
Cranberry	2.0–4.5	90–100	12–16 weeks

TABLE 3.4 *(Continued)*

Crops	Temperature Range (°C)	Relative Humidity	Storage Time
Current	–0.5–0.0	90–95	1–2 weeks
Kiwifruit	–0.5–0.0	90–95	8–16 weeks
Nectarine	–0.5–0.0	85–90	1–6 weeks
Nuts (almond, pistachio, walnut)	0.0	65–70	16–96 weeks
Peach	–0.5–0.0	90–95	2–6 weeks
Plum	–2.0–0.0	90–95	1–7 weeks
Pear	–2.0–0.0	85–95	8–26 weeks
Persimmon	–1.0–0.0	90–95	12–16 weeks
Raspberry	–0.5–0.0	90–95	2–3 days
Strawberry	–0.5–0.0	90–95	5–14 days
	Tropical/Subtropical Fruits		
Acid lime	7–10	85–90	4–10 weeks
Avocado	5–13	85–95	2–4 weeks
Banana	13–15	85–90	4–21 days
BER	8–10°C	75–80%	5–6 weeks
Carambola	10–15	85–90	5 weeks
Custard apple	10–15	85–90	2–3 weeks
Grapes	–1.0–0°C	90–95%	4–8 weeks
Guava	7–10	85–90	2–3 weeks
Indian gooseberry	10°C	40%	8–10 days
Jackfruit	13	85–90	2–6 week
Lime	8–10	85–90	6–8 weeks
Litchi	2–3	90–95	3–5 week
Mandarin	6–8	85–90	3–8 week
Mango	12–13	85–90	2–3 week
Mangosteen	13	85–90	2–4 weeks
Papaya	8–13	85–90	1–3 week
Passion fruit	8–10	85–90	2–3 weeks
Pineapple	7–13	85–90	20–24 weeks
Pomegranate	0	90–95	8–12 weeks
Rambutan	12–13	85–90	1–3 weeks
Sapota	16–20	85–90	2–3 weeks
Sweet orange	5–7°C	90–95%	10–12 weeks

TABLE 3.4 *(Continued)*

Crops	Temperature Range (°C)	Relative Humidity	Storage Time
	Vegetables		
Cabbage	0	98–100	3–6 weeks
Cauliflower	0	98–100	3–4 weeks
Broccoli	0	98–100	2 weeks
Brinjal	12	90–95	1 week
Tomato	13–15	90–95	1 week
Potato	5–13	90–95	8–10 months
Chili	7–13	90–95	2–3 weeks
Beans	4–7	90–95	7–10 days
Peas	4–10 °C	90–95%	1–2 weeks
Cucumber	10–13	90–95	2 weeks
Sweet corn	0	90–95	5–8 days
Pumpkin	10–13	50–70	2–3 months
Onion	0	65–70	6–8 months
Garlic	0	65–70	5–6 months
Carrot	0	98–100	4–6 weeks
Reddish	0	90–95	2–4 months
Yam	16	70–80	20–24 weeks
Sweet potato	13–15	85–90	4–7 weeks
Spinach	0	95–98	10–14 days
Lettuce	0	98–100	2–3 weeks
Okra	7–10	90–95	7–10 days
Mushroom	0	95–98	3–4 days

3.4.2 HIGH-PRESSURE PROCESSING (HPP)

High-pressure processing (HPP) is a well-established non-thermal technology for ensuring microbial safety and nutritional quality of foods. Inactivation of deleterious enzymes is achieved through the application of high-pressure technology. The most important feature of this technology is that it has minimal effect on the flavor and nutritional value of foods at low temperature. High pressure processed foods demonstrate better stability in fruit quality during refrigeration as compared to thermally processed ones (Tewari et al., 2016).

3.4.3 STERILIZATION TECHNOLOGY

Microbial contamination of fruit and vegetables can occur at different stages of handling, which may affect consumers' health. The increasing demand for safe and healthy food has paved the way for sterilization technology to keep the produce free from microbes as well as minimize the use of chemicals in preservation. There many approaches for food sanitation, and among them, radiation technology and ozone storage are more common. The appropriate use of radiation can extend the shelf life of produce without being stored at low temperature. Radiation technology delays the ripening of F&V and limits the deterioration of quality of stored tuber and bulb crops by preventing postharvest sprouting Wang and Chao (2003). The technology mainly uses γ-rays and X-rays to disinfect, sterilize, and kill microbes by ionizing and producing free radicals. Radiation treatment keeps the produce sterilized without affecting its original flavor and nutritive value. Moreover, the treatment also prolongs the shelf life of produce at normal temperature. The use of 0.15 to 0.3 kGy ray to irradiate fruit and vegetable causes disinfection without affecting their quality. UV radiation technology (UVRT) is also commonly used to preserve the horticultural produce. The use of non-ionizing, germicidal UV light could be effective to disinfect F&V as a whole or as fresh-cut products. The exposure of bell pepper, zucchini squash, lettuce, potato, tomato, apple, pear, and strawberry to UV light reduced microbial activity and deterioration during subsequent storage at low temperature. The UVRT stimulates respiration rate without affecting ethylene production (Erkang, 2008). Ozone seems to have beneficial effects in extending the shelf life of fresh commodities such as broccoli, cucumber, apples, grapes, oranges, pears, raspberries, and strawberries by reducing microbial populations and through ethylene oxidation. Treatments on apples with ozone resulted in a reduction of weight loss and spoilage. Vegetables treated with ozone during storage showed a considerable decrease in mold and bacterial counts without causing any change in their chemical composition and sensory quality (Karaka, 2007; Horvitz and Cantalajo, 2014).

3.5 CONCLUSION

Packaging and storage has become an integral part of the horticulture production system due to its significance in improving shelf life and minimizing postharvest loss. The future of packaging is one in which the package will increasingly operate as a smart system incorporating both smart and conventional materials, adding value and benefits across the packaging supply chain. Packaging should be low cost and environmentally safe. In order to ensure better marketability higher package quantity is needed for fresh F&V active packaging and intelligent packaging are emerging and exciting areas of food preservation technique, which can confer many preservation benefits on a wide range of foods. The aim of active packaging is to match the properties of the package to the more critical requirements of the food. Adoption of some of these methods will require changes in attitude to packaging and a willingness to address regulatory issues where chemical effects are used. Application of new emerging technologies offers greater potential in horticulture sector. The intelligent system is an effective approach not only to maintain the quality or value of a product but also to provide information about produce quality to consumers.

KEYWORDS

- **controlled atmosphere packaging**
- **hazard analysis and critical control points**
- **high-pressure processing**
- **modified atmosphere packaging**
- **potassium permanganate**
- **quality analysis and critical control points**
- **radiofrequency identification**
- **scavenger**

REFERENCES

Alvarez-Hernandez, M. H., Artes-Hernandez, F., Avalos-Belmontes, F., Castillo-Campohermoso, M. A., Contreras-Esquivel, J. C., Ventura-Sobrevilla, J. M., & Martínez-Hernández, G. B., (2018). Current scenario of adsorbent materials used in ethylene scavenging systems to extend fruit and vegetable postharvest life. *Food Bioprocess Technol., 11*, 511–525.

Arvanitoyannis, I. S., & Oikonomou, G., (2012). Active and intelligent packaging. In: Arvanitoyannis, I. S., (ed.) *Modified Atmosphere and Active Packaging Technologies* (pp. 628–54). Boca Raton, Florida, U.S.A.: CRC Press.

ASSOCHAM, (2013). *Post-Harvest Losses*. The Associated Chambers of Commerce and Industry of India, New Delhi.

Ayele, L., Tsadik, W. K., Abegaz, K., & Yetneberk, S., (2012). Postharvest ripening and shelf life of mango (*Mangifera indica* L.) fruit as Influenced by 1-Methylcyclopropene and Polyethylene Packaging. *Ethiop. J. Agric. Sci., 22*, 26–44.

Azene, M., Workneh, T. S., & Woldetsadik, K., (2014). Effect of packaging materials and storage environment on postharvest quality of papaya fruit. *Journal of Food Science and Technology, 51,* 1041–1055.

Bhattarai, B. P., & Shah, R., (2017). Effect of different packaging materials on postharvest status of mandarin (*Citrus reticuleta* Blanco). *Journal of Horticulture, 4*(4), 218. doi: 10.4172/2376-0354.1000218.

Cruz, R. S., Camilloto, G. P., & Santos, P. A. C., (2012). Oxygen scavengers: An approach on food preservation. In: Eissa, A. A., (ed.), *Structure and Function of Food Engineering* (pp. 21–42). Intech Open. http://dx.doi.org/10.5772/48453.

Damaj, Z., Naveau, A., Dupont, L., Hénon, E., Rogez, G., & Guillon, E., (2009). Co(II) (L-proline)2(H2O)2 solid complex: Characterization, magnetic properties, and DFT computations. Preliminary studies of its use as oxygen scavenger in packaging films. *Inorg. Chem. Commun., 12*, 17–20.

Day, B. P. F., (2003). Active packaging. In: Coles, R., McDowell, D., & Kirwan, M., (eds.), *Food Packaging Technologies* (pp. 282–302). Boca Raton, FL, USA: CRC Press.

Dhall, R. K., Sharma, S. R., & Mahajan, B. V. C., (2010). Effect of packaging on storage life and quality of cauliflower stored at low temperature. *Journal of Food Science and Technology, 47*(1), 132–135.

EFSA, (2009). www.efsa.europa.eu (accessed on 19 December 2020).

Erkan, M., Wang, S. Y., & Wang, C. Y., (2008). Effect of UV treatment on antioxidant capacity, antioxidant enzyme activity, and decay in strawberry fruit. *Postharvest Biology and Technology, 48*, 163–171.

FAO, (2011). *The Role of Women in Agriculture*. ESA Working Paper No. 11-02. Agricultural Development Economics Division, The Food and Agriculture Organization of the United Nations. Available at: www.fao.org/economic/esa (accessed on 19 December 2020).

Horvitz, S., & Cantalejo, M. J., (2014). Application of ozone for the postharvest treatment of fruits and vegetables. *Crit. Rev. Food Sci. Nutr., 54*, 312–339.

Hynniewta, L. R., Banik, A. K., & Singh, L. J., (2017). Study on the effects of packaging and storage of himsagar mango. *International Journal of Current Microbiology and Applied Sciences, 6*, 1044–1048.

Ji, L., Pang, J., Li, S., Xiong, B., & Cai, L., (2012). Application of new physical storage technology in the fruit and vegetable industry. *African Journal of Biotechnology, 11*, 6718–6722.

Kader, A. A., & Barrett, D. M., (2005). Classification, composition of fruits, and postharvest maintenance of quality. In: Barrett, D. M., Somogyi, L., & Ramaswamy, H., (eds.), *Processing Fruits Science and Technology* (2nd edn., pp. 3–21). CRC Press.

Kader, A. A., & Saltveit, M. E., (2005). Respiration and gas exchange. In: Barrett, D. M., Somogyi, L., & Ramaswamy, H., (eds.), *Postharvest Physiology and Pathology of Vegetables* (2nd edn., pp. 7–32). CRC Press.

Kader, A. A., (2002). Pre- and postharvest factors affecting fresh produce quality, nutritional value, and implications for human health. In: *Proceedings of the International Congress Food Production and the Quality of Life* (pp. 109–119). Italy.

Kader, A. A., Zagory, D., & Kerbel, E. L., (1989). Modified atmosphere packaging of fruits and vegetables. *Critical Reviews in Food Science and Nutrition, 28*(1), 1–30.

Karaca, H., & Velioglu, S. Y., (2007). Ozone applications in fruit and vegetable processing. *Food Rev. Int., 23*, 91–106.

Kudachikar, V. B., Kulkarni, S. G., & Prakash, M. N. K., (2011). Effect of modified atmosphere packaging on quality and shelf life of 'Robusta' banana (*Musa* sp.) stored at low temperature. *Journal of Food Science and Technology, 48*, 319–324.

Kusumaningrum, D., Lee, S. H., Lee, W. H., Mo, C., & Cho, B. K., (2015). A review of technologies to prolong the shelf life of fresh tropical fruits in Southeast Asia. *J. of Biosystems Eng., 40*(4), 345–358.

Manurakchinakorn, S., Nuymak, P., & Issarakraisila, M., (2014). Enhanced chilling tolerance in heat-treated mangosteen. *International Food Research Journal, 21*, 173–180.

Matche, R. S., (2013). Packaging aspects of fruits and vegetables. In: *Plastics in Food Packaging*, 117–132.

Molla, M. M., Islam, M. N., Muqit, M. A., Ara, K. A., & Talukder, M. A. H., (2011). Increasing shelf life and maintaining quality of mango by postharvest treatments and packaging technique. *Journal of Ornamental and Horticultural Plants, 1*(2), 73–84.

Phong, N. V., & Nhung, D. T. C., (2016). Effects of microperforated polypropylene film packaging on mangosteen fruits quality at low temperature storage. *Journal of Experimental Biology and Agricultural Sciences, 4*, 706–713.

Rathore, H. A., (2009). Effect of polythene packaging and coating having fungicide, ethylene absorbent and antiripening agent on the overall physic-chemical composition of Chausa White variety of mango at ambient temperature during storage. *Pakistan Journal of Nutrition, 8*(9), 1356–1362.

Rene, T., (2001). Preservation of archives in tropical climate. In: *The International Council on Archives, the National Archives of the Republic of Indonesia and the National Archives of the Netherlands on the Occasion of the International Conference*. Jakarta.

Rooney, M. L., (1995). *Active Packaging*. Blackie Academic & Professional, New York.

Scetar, M., Kurek, M., & Galic, K., (2010). Trends in fruit and vegetable packaging: A review. *Croatian Journal of Food Technology, Biotechnology and Nutrition, 5*, 69–86.

Sisler, E. C., & Serek, M., (2003). Compounds interacting with the ethylene receptor in plants. *Plant Biol., 5*, 473–480.

Skog, L. J., & Chu, C. L. (2001). Effect of ozone on qualities of fruits and vegetables in cold storage. *Canadian Journal of Plant Science, 81*, 773–778.

Smith, A. W. J., Poulston, S., & Rowsell, L., (2009). A new palladium-based ethylene scavenger to control ethylene-induced ripening of climacteric fruit. *Platinum Metals Rev., 53*(3), 112–122.

Spadaro, D., Vola, R., Piano, S., & Gullino, M. L., (2002). Mechanisms of action and efficacy of four isolates of the yeast *Metschnikowia pulcherrima* active against postharvest pathogens on apples. *Postharvest Biology Technology, 24*(2), 123–134.

Tewari, S., Sehrawat, R., Nema, P. K., & Kaur, B L., (2017). Preservation effect of high-pressure processing on ascorbic acid of fruits and vegetables: A review. *Journal of Food Biochemistry, 41*, e12319.

Thakur, A. K., Kumar, R., Shambhu, V. B., & Singh, I. S., (2017). Effectiveness of shrink-wrap packaging on extending the shelf life of apple. *Int. J. Curr. Microbiol. App. Sci., 6*, 2365–2374.

Wang, J., & Chao, Y., (2003). Effect of gamma irradiation on quality of dried potato. *Rad. Phys. Chem., 66*, 293–297.

Warton, M. A., Wills, R. B. H., & Ku, V. V. V., (2000). Ethylene levels associated with fruit and vegetables during marketing. *Aust. J. Expt. Agric., 40*, 485–490.

Wong, D. E., Andler, S. M., Lincoln, C., Coat, J. et al., (2017). Oxygen scavenging polymer coating prepared by hydrophobic modification of glucose oxidase. *Technol. Res., 14*, 489. https://doi.org/10.1007/s11998-016-9865-6.

Wu, C. T., (2010). An overview of postharvest biology and technology of fruits and vegetables. In: *Workshop on Technology on Reducing Postharvest Losses and Maintaining Quality of Fruits and Vegetables* (pp. 2–11).

Yildirim, S., Rocker, B., Ruegg, N., & Lohwasser, W., (2015). Development of palladium-based oxygen scavenger: Optimization of substrate and palladium layer thickness. *Packag. Technol. Sci., 28*(8), 710–718.

Zerdin, K., Rooney, M. L., & Vermuë, J., (2003). The vitamin C content of orange juice packed in an oxygen scavenger material. *Food Chem., 82*, 387–395.

Zhu, S. J., (2006). Non-chemical approaches to decay control in postharvest fruit. In: Noureddine, B., & Norio, S., (eds.), *Advances in Postharvest Technologies for Horticultural Crops* (pp. 297–373). Research Signpost, Trivandrum, India.

CHAPTER 4

BIODEGRADABLE PACKAGING: A SUSTAINABLE APPROACH FOR PACKAGING OF FRUITS AND VEGETABLES

ABHISHEK DUTT TRIPATHI,[1] ARPIT SHRIVASTAVA,[1] KAMLESH KUMAR MAURYA,[1] DIVYA KUMARI KESHARI,[1] and TANWEER ALAM[2]

[1]*Centre of Food Science and Technology Institute of Agricultural Sciences, Banaras Hindu University, Varanasi, Uttar Pradesh, India, E-mail: abhi_itbhu80@rediffmail.com (A. D. Tripathi)*

[2]*Indian Institute of Packaging, New Delhi, India*

ABSTRACT

This chapter offers an overview of the production, usage and current prospects of biodegradable packaging, The types of biopolymers, the properties of biodegradable packaging, the notions of biodegradability the benefits and drawbacks of biotechnological production and packaging. Conventional food packaging materials have shown shortcomings in expression of their environmental pollution impact and in their manufacturing requirements for non-renewable resources, the necessity for alternate packaging materials and packaging set-ups is now needed additional than ever. The biodegradability of packaging material is one of the major challenges which can be overcome by using biodegradable packaging materials. Growing interest in replacing goods relying on petroleum with natural cheap and renewable materials are essential for forthcoming sustainable growth. A number of bio-based materials and their innovative applications in food-related packaging have gained much attention over the past several years. The majority of bio based packages showed appropriate barriers for humidity and gases to act as food packaging material and packaging of food items with a long shelf life is possible. These new materials for packaging of fruits and vegetables are discussed in this chapter.

4.1 INTRODUCTION

Food packaging is one of the most protruding parameters and protects the food item from contamination and maintains the quality of food freshness, increases shelf life of a food product. The principal role of food packaging is not only protecting the food material from outside but also maintains the food quality, ingredient, nutritional information, and advertisement of product. Traditionally glass, metal, paper, paperboards, and plastics are used as a food packaging materials. Different packaging materials used in food packaging, generally polypropylene (PP) based plastics are more commonly used, but they pose environmental threats as they emit many harmful gases upon emission and also liberates toxic substances affecting the human health. Conventional food packaging materials have shown shortcomings in expression of their environmental pollution impact and in their manufacturing requirements for non-renewable resources, the necessity for alternate packaging materials and packaging set-ups is now needed additional than ever. The biodegradability of packaging material is one of the major challenges which can be overcome by using biodegradable packaging materials. Biodegradable apply to the efficiency of materials to infringe down and revert to naturalness within a narrow delay after removal naturally in a year or less. Development of edible/biodegradable films/coatings for efficient food packaging has conceived noteworthy interest in recent ages due to their potential to reduce and/or substitute conventional, non-biodegradable plastics. Various bio-based materials and their imaginative applications in food-related packaging have increased plentiful consideration in the course of recent years. These ongoing materials contain starch, cellulose, and those start from forms, including microbial fermentation. Bioplastic improvement endeavors have focused fundamentally upon starch, which is an inexhaustible and widely present crude material. As far as a financial state, starch is less expensive than oil and has been utilized in a few techniques for arrangement compostable plastics. Corn is the crude wellspring of starch for bioplastics, albeit increasingly novel no matter how you look at it explores are assessing its strong application in bioplastics for starches from potato, wheat, rice, grain, oat, and soy sources. Some potential employments of these materials are the safety of fruit and vegetable foods. Packaging fresh food material is one of the most important not only producer but also consumer. Container, crates, sacks cases, hampers mass receptacles, and palletized containers are appropriate compartments for taking care of, shipping, and promoting fresh produce. In the green part of the world, India is the biggest maker of organic products (46 million tons) with a worldwide portion of over 10% and the second-biggest maker of vegetables (80 million tons) with a worldwide portion of over 15%. Regardless of every one of these accomplishments, around 20 to 30% of the produce is lost yearly because of nonattendance of good framework and less utilization of current post-reap innovations. Generally, food organizations utilize polymeric films (polyethylene (Pe), PP, polystyrene (PS)) to packaging fresh fruits and vegetables (F&V) as a result of their huge accessibility at moderately minimal effort and their great mechanical exhibition, great hindrance to oxygen, carbon dioxide. The present work centers on the diverse packaging procedures for fresh-cut F&V. Specifically, the potential uses of biodegradable materials and eatable coatings have been characterized, with their consequences for the nature of these items.

4.2 BIODEGRADABLE PACKAGING

Biodegradable apply to the efficiency of materials to infringe down and revert to naturalness within a narrow delay after removal, typically in a year or less. New biodegradable polymer mixes have been created to improve the debasement of the last item. The utilization of biopolymers depends on sustainable assets and adds to material cycling that is practically equivalent to the common biogeochemical cycles in nature (Buccin et al., 2005).

Biopolymers, as it can be said instead of biodegradable plastics, may be naturally occurring materials. Most of the raw material to biodegradable materials comes from the nature; green plants, bacteria, or even from the eatable vegetables. The natural synthesis is a very complex process and it is not yet practical as a complete production for commodity plastics. In the food industry the usage of biodegradable materials are still in ascent phase and only few examples are available in this field. Polymer materials are strong, non-metallic mixes of high atomic weight. They are included rehashing macromolecules, and have changing attributes relying on their composition. Every macromolecule that involves a polymeric material is known as a mer unit. A solitary mer is known as a monomer, while rehashing mer units are known as polymers. New biodegradable polymer mixes have been created to improve the degradation of the last item.

The use of biopolymers is based on renewable resources and contributes to material cycling that is analogous to the natural biogeochemical cycles in nature. In addition to the above environmental issues, food packaging has been impacted by notable changes in food distribution, including globalization of the food supply, consumer trends for more fresh and convenient foods, as well a desire for safer and better-quality foods. Given these and previously mentioned issues, consumers are demanding that food packaging materials be more natural, disposable, potentially biodegradable, as well as recyclable (López-Rubio et al., 2004). "There has been a growing interest and effort over the last few years in the development of novel food packaging concepts, which can play a proactive role regarding product preservation, shelf-life extension, and even improvement".

4.2.1 TYPES OF BIODEGRADABLE PACKAGING

It is basic to separate between the different kinds of biodegradable plastic, as their cost and uses are very unique.

The two driving sorts are oxo-biodegradable and hydro-biodegradable. In the two conditions, degeneration continues with a diagnostic procedure (oxidation and hydrolysis individually), joined by an organic procedure. The two kinds discharge CO_2 as they degrade, but mostly hydro-biodegradable can transmit methane which is very harmful to our environment. The two kinds are compostable; however, just oxo-biodegradable can be monetarily reused. Hydro-biodegradable is significantly more costly than oxo-biodegradable.

4.2.1.1 OXO-BIODEGRADABLE PLASTIC

Plastic created from this ongoing innovation gets deteriorated by a procedure of OXO-degradation. The technology is based on a very mean amount of prodegradant additive being inducted into the manufacturing progress; thereby alter the behavior of the plastic. Degradation begins after the programmed service life is completed (as measured by the additive formulation) after the requirement of product is over. This technology produces such a product that doesn't need additional cost and is identical to conventional plastic products in terms of machinery and workforce. The plastic used up by bacteria and fungi will just not fragment after the additive but also will reduce to molecular level, which authorizes living micro-organisms access to the carbon and hydrogen. It is thus "biodegradable." Till the material has biodegraded to nothing more than CO_2, water, and humus this process lasts, fragments of petro-polymers are not left in the soil. Oxo-biodegradable plastic is usually passed in all the ecotoxicity tests, including seed germination, plant growth, and organism survival (daphnia, earthworms).

Naptha, a by-product of oil refining are utilized for Oxo-biodegradable plastics production, and oil is obviously a limited resource. Though, this by-product ascends cause the world desires fuels and oils for engines and would arise whether or not the by-product were used to make plastic goods. If the oil will be left under the ground, carbon dioxide will unavoidably be released, but until other fuels and lubricants have been developed for engines, it makes good eco-friendly sense to use the by-product, as an alternative of worsening it by "flare-off" at the refinery and consuming threatened agricultural resources to make plastics.

4.2.1.2 HYDRO-BIODEGRADABLE PLASTICS

Hydrolysis presents hydro-biodegradation out of this gathering a few plastics have a high starch substance, and it is once in a while said this legitimizes the benefit that they are produced using sustainable assets. However, many of them contains up to half of manufactured plastic derivatives of oil, and others (for example, some aliphatic polyesters) are totally based on oil-derived intermediates in the production of hydro-biodegradable plastics mostly genetically changed crops are used. Hydro-biodegradable plastics are not evidently "sustainable" as the procedure of their assembling generally from crops is itself a vital handler of petroleum product vitality and a maker along these greenhouse gases harming substances. Petroleum derivatives are singed in the autoclaves used to mature and polymerize material incorporated from biochemically created intermediates (for example, polylactic corrosive from starches and so on); and by the agricultural machinery and street vehicles utilized; likewise, by fabricate and transport of composts and pesticides. They are infrequently marked as through from "non-food" crops, however are in certainty ordinarily produced using food crops.

4.2.1.3 PHOTO-DEGRADABLE PLASTICS

These respond to ultra-violet light, however, except if they are likewise oxo-biodegradable they won't degrade in a landfill, a sewer, or other dull condition, or if vigorously overprinted.

4.2.2 CLASSIFICATION OF BIODEGRADABLE PACKAGING

4.2.2.1 POLYMERS DIRECTLY EXTRACTED/REMOVED FROM NATURAL MATERIALS

4.2.2.1.1 Polysaccharides

Depending on the process activity and on the source, biopolymers can feature properties similarly with traditional ones.

These materials can be:

i. Polymers with direct extraction from biomasses like proteins, lipids, polysaccharides, etc.;
ii. Materials polymeric in nature produced by microorganisms and bacteria such as polyhydroxyalkanoates (Popa et al., 2011).

Characteristic related to composability is chief for biopolymer materials, as composting permits dumping of the packages in the soil, if compared with recycling, which is more energy-efficient. In biological degradation, inorganic compounds such as water, carbon dioxide is formed without residues toxic in nature.

Effective gas barrier properties are shown by polysaccharides though they are extremely hydrophilic and high water vapor permeability is shown if compared with commercial plastic films. The starch and starch derivates, cellulose derivates, alginate, carrageenan, chitosan, pectin, and several gums are chief polysaccharides that can be included in edible coating arrangements. Difference can be based at molecular level among polysaccharides due to their molecular weight, degree of branching, conformation, electrical charge, and hydrophobicity.

4.2.2.1.2 Starch

Starch is a complex polysaccharide and it is hydrocolloid in nature and well known among biopolymers. It is a cheap polysaccharide, profusely available and one of the low-priced biodegradable polymers. Starch is formed in the form of granules, hydrophilic in nature and gathered through agricultural plants. Potatoes, corn, wheat, and rice are its main source from which it is extracted. Composition of starch is amylose (poly-1,4-D-glucopyranoside), a linear and crystalline polymer and amylopectin (Poly-1, 4-D-glucopyranoside and -1,6-D-glucopyranoside), a branched and amorphous polymer. Starch contains different amount of amylopectin and amylose ranging from about 80–90% amylopectin and 10–20% amylase which depend on source. Amylose forms a helical structure in water and is soluble, and the comparative quantities and molar masses of amylose and amylopectin differ with the starch source, producing materials which differ in their mechanical and biodegradable properties and. Starch is typically used as a thermoplastic. Through destructuration, it is plasticized with precise amounts of water or plasticizers after heating, it is extruded. Thermoplastic starch (TPS) is highly sensitive to humidity. Thermal properties of TPS are influenced by the amount of water rather than starch molecular weight. TPS produced is almost amorphous. By acetal link enzymes starch biodegraded through hydrolysis. The -1, 4 link is attacked by amylases while glucosidases attack the -1, 6 link. The degradation products are nonpoisonous (Figure 4.1).

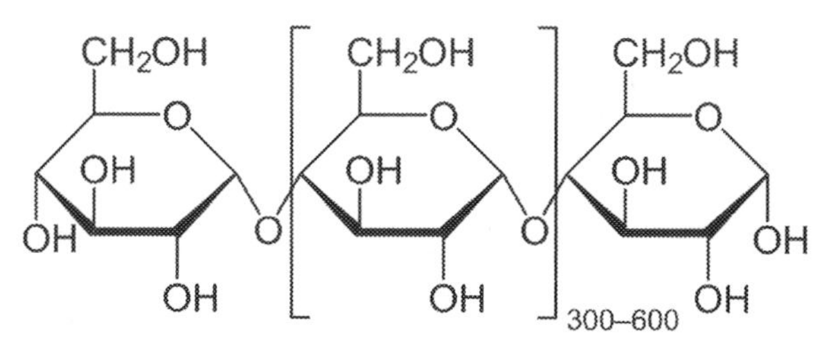

FIGURE 4.1 Starch structure.

4.2.2.1.3 Chitin

It is a biopolymer which is second most abundant among natural polymer. It is a linear copolymer of N-acetyl-glucosamine and N-glucosamine with -1,4 linkage. These units are haphazardly or block circulated all through the biopolymer chain, which depends on the process used for obtaining biopolymer. Shells of crabs, shrimp, crawfish, and insects are the usual source of Chitin. It could be considered as amino cellulose. Current researches suggest that an alternative source of chitin can be the cultivation of fungi (Figure 4.2).

4.2.2.1.4 Chitosan

Chitin by partial alkaline N-deacetylation produces chitosan. In chitosan, glucosamine units are prime component. Glucosamine to acetyl glucosamine ratio is described as the degree of deacetylation. This degree may extend from 30% to 100% contingent upon the arrangement strategy, and it influences the crystallinity, surface vitality, and debasement pace of chitosan. Biodegradation of cellulose continues by enzymatic oxidation, with peroxidase emitted by growths. Cellulose can likewise be debased by microorganisms. With respect to starch degradation items are non-poisonous (Vroman and Tighzert, 2009).

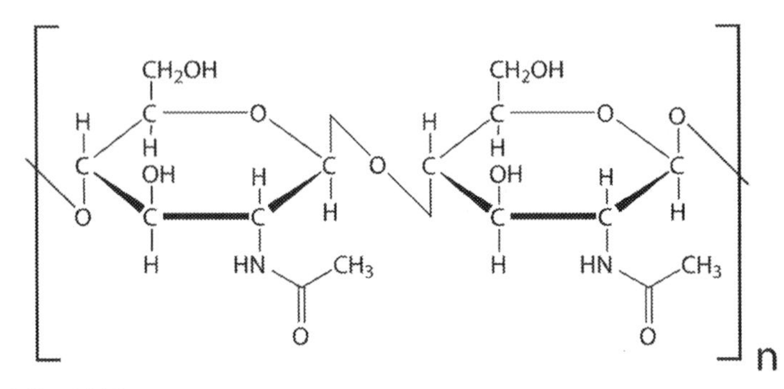

FIGURE 4.2 Chitin structure.

4.2.2.1.5 *Cellulose*

Cellulose is a polysaccharide produced by plant and widely used in various industries. Cellulose structure is a linear polymer with very long macromolecular chains of one repeating unit, cellobiose. Properties of cellulose are crystalline, infusible, and insoluble in all organic solvents. Cellulose esters are altered polysaccharides, and wide-ranging transformation can be obtained by altering the mechanical properties and biodegradation properties decreases with an increase in degree of substitution (Figure 4.3).

4.2.2.1.6 *Proteins*

Corn zein, wheat gluten (WG), soy protein, whey protein, casein, collagen/gelatin, pea protein, rice bran protein, cottonseed protein, peanut protein, and keratin are proteins that have been acknowledged for their ability of forming edible films and coatings including casein based edible coatings are striking for food applications because of high nutritional quality, whey proteins have been presented to extraordinary examination over the past period or so with the expansion of plasticizer, heat-denatured whey proteins produce straightforward and adaptable water-based consumable coatings with great oxygen, smell, and oil obstruction properties at low relative mugginess. Be that as it may, the hydrophilic idea of whey protein coatings makes them be less successful as moisture barriers.

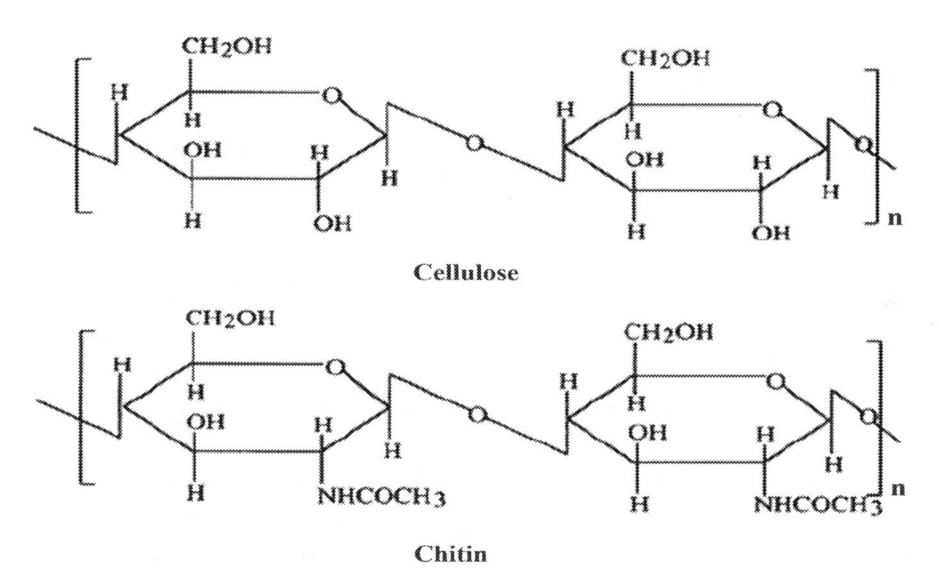

FIGURE 4.3 Cellulose structure.

4.2.2.1.7 Lipids

Edible lipids used to create consumable coatings are: beeswax, candelilla wax, carnauba wax, triglycerides, acetylated monoglycerides, unsaturated fats, greasy alcohols, and sucrose unsaturated fat esters. Lipid coatings and movies are for the most part utilized for their hydrophobic properties, speaking to an average obstruction to dampness misfortune. Lipids structures thicker and progressively fragile movies because of their hydrophobic trademark, therefore they should be related with film framing specialists, for example, proteins or cellulose subordinates.

4.2.2.2 BIODEGRADABLE POLYMERS DERIVED FROM PETROLEUM RESOURCES

These polymers are manufactured with hydrolyzable capacities, for example, ester, amide, and urethane, or polymers with carbon spines, in which are included added substances, for example, cell reinforcements. Synthesis, properties, and biodegradability of the fundamental classes and new groups of manufactured polymers are talked about underneath:

4.2.2.2.1 Polymers with Additives

The conventional polymers mostly are unaffected to degradation and are derived from petroleum resources. Biodegradation is facilitated, by the introduction of additives. Polyolefins are degraded by overview of antioxidants into the polymer chains. Degradation is accomplished by reaction of Antioxidants under UV by photo-oxidation. Be that as it may, the biodegradability of such frameworks is as yet questionable.

4.2.2.2.2 Synthetic Polymers with Hydrolysable Backbones

Under specific conditions, polymers with hydrolyzable backbones are vulnerable to biodegradation. Polymers developed with these polymers have properties including polyesters, polyamides, polyurethanes, polyurea's, poly (amide-enamine), and polyanhydrides.

4.2.2.2.3 Aliphatic Polyesters

It is one the most extensively considered class of biodegradable polymers; because of its synthetic flexibility their significant variety and a large variation of monomers can be used. Numerous ways for the development of synthetic polyesters exists and have been recently reviewed Examples are poly (glycolic acid) or poly (lactic acid). Poly (alkene dicarboxylate) signifies the second class. Aliphatic polyesters are synthesis by polycondensation of diols and dicarboxylic acids. Examples are poly (butylene succinate) and poly (ethylene succinate) (Vroman and Tighzert, 2009).

4.3 POLYLACTIC ACID (PLA)

The monomeric building block of PLA, is formed by from vegetable sources (for example, corn, wheat, or rice) using either bacterial fermentation or petrochemical route by conversion of sugar or starch. Lactic acid exists as two optical isomers, l-, and d-lactic acid. It is a helpful polymer, recyclable, and compostable, with high straight forwardness, high atomic weight, great processability and water solubility resistance (Figures 4.4 and 4.5).

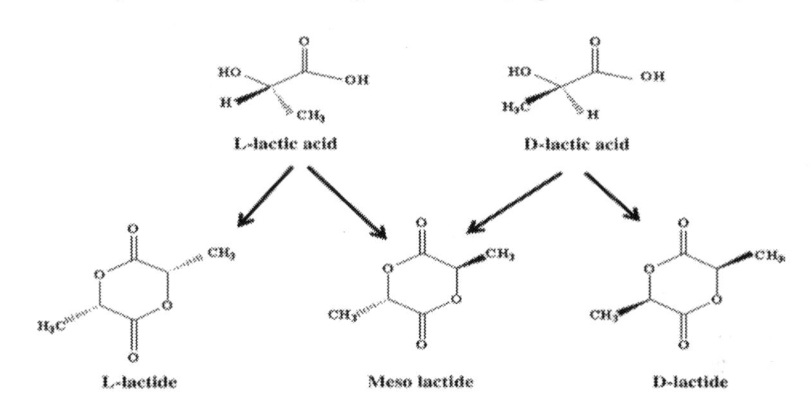

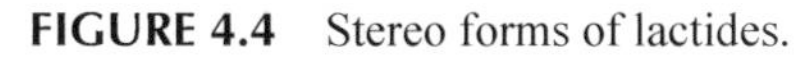

FIGURE 4.4 Stereo forms of lactides.

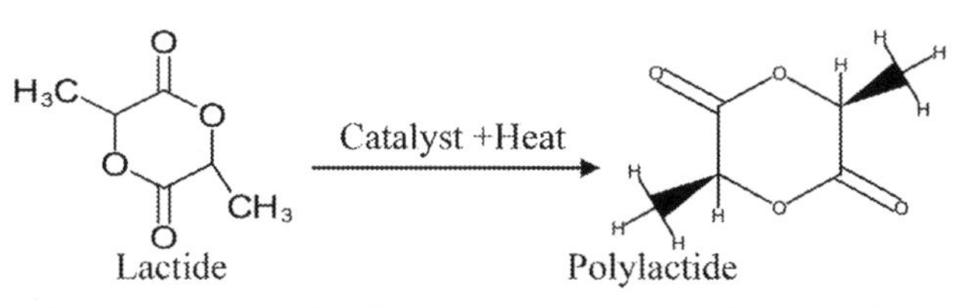

FIGURE 4.5 Schematic view of polymerization reaction of polylactide from lactide, a dimer of lactic acid.

Source: Manually made by Pawar and Purwar (2013).

The chief microbial sources of lactic acid are lactic acid bacteria (LAB) and some filamentous fungi. L-isomer are predominantly produce by the microorganism such as *Lactobacilli amylophilus, Lactobacilli bavaricus, Lactobacilli casei, Lactobacilli salivarius.* Microbial strains such as *Lactobacilli delbrueckii, Lactobacilli acidophilus, Lactobacilli jensenii* yield the D-isomer or mixture of both. For microbial production of lactic acid, carbohydrate-rich sources are needed which can be either sugar in pure form such as glucose, sucrose, lactose, etc., or sugar-containing materials such as molasses, cassava, starchy materials from potato, tapioca, wheat, whey, etc., (Tables 4.1 and 4.2).

4.3.1 DEGRADATION OF PLA

Cleavage of main chains or side chains of macromolecules of PLA leads to the occurrence of polymer degradation. Naturally, the entire process which induces the polymer degradation such as thermal activation, hydrolysis, biological activity, in many cases, could also be referred to as environmental degradation (Muller, 2008). Environmental factors not only causes polymer degradation but also have a major influence on the microbial population and on their activity. Degradation of PLA shows its dependency on several factors such as molecular weight, crystallinity, purity, temperature, pH, presence of terminal carboxyl or hydroxyl groups, additives acting catalytically which may include enzymes, bacteria or inorganic fillers (Park and Xanthos, 2009).

TABLE 4.1 The Substrate and Microorganism Used to Produce Lactic Acid and its Yield

Substrate	Micro Organism	Lactic Acid Yield
Wheat and rice bran	*Lactobacillus sp.*	129 g/l
Corn cob	*Rhizopus sp. MK-96-1196*	90 g/l
Pretreated wood	*Lactobacillus delbrueckii*	48–62 g/l
Cellulose	*Lactobacillus coryniformis ssp. torques*	0.89 g/g
Barley	*Lactobacillus caseiNRRLB-441*	0.87–0.98 g/g
Cassava bagasse	*L. delbrueckii NCIM 2025, L. casei*	0.9–0.98 g/g
Wheat starch	*Lactococcus lactis ssp. lactis ATCC 19435*	0.77–1 g/g
Whole wheat	*Lactococcus lactis, Lactobacillus delbrueckii*	0.93–0.95 G/G
Potato starch	*Rhizopus oryzae, R. arrhizuso*	0.87–0.97 g/g
Corn starch	*L. amylovorous NRRL B-4542*	0.935 g/g
Corn, rice, wheat starches	*Lactobacillus amylovorous ATCC 33620*	<0.70 g/g

TABLE 4.2 Properties of PLA

Property	Value
Technical name	Polylactic acid
Chemical formula	$(C_3H_4O_2)n$
Melting temperature	157–170°C
Heat deflection temperature	49–52°C
Tensile strength	61–66 MPa
Flexural strength	48–110 MPa
Specific gravity	1.24
Shrink rate	0.37–0.41%
Density	1.210–1.430 g/cm^3
Glass transition temperature	60–65°C
Crystallinity	37%
Solubility	Chlorinated solvents, hot benzene, tetrahydrofuran, and dioxane

Copyrolytic technique offers an alternative way to waste treatment and also act as an upgrading step during the pyrolysis of biomass (Gang and Almin, 2008). Hydrolysis treatment applied to PLA at temperature 180°C–350°C for 30 mins, leads to the formation of L-lactic acid which can be again recycled back to the monomer (Auras et al., 2004). Controlled composting environment said to breakdown PLA into carbon dioxide and water in less than 90 days (Ghorpade et al., 2001). Degradation of PLA soft films was observed within 3 weeks while the PLA rope samples and band samples shows its degradation within 6 weeks. Hence, all the PLA products get rapidly degraded under composting (Kimura et al., 2002). Biodegradation of PLA can also be achieved by esterases, proteases, lipases secreted from microorganisms. Among many commercial protease enzymes for degradation of PLA, Savinase 16 L from *Bacillus lentis* and Protin A from *B. subtiis* showed higher degradation activity (Oda et al., 2000). Actinomycete strain can degrade PLA which was taxonomically similar to Amycolatopsis (Tokiwa and Jarerat, 2004). A fungal strain, *Tritirachium album* can also lead to the PLA degradation (Jarerat and Tokiwa, 2001, 2003). PLA degradation rate could also be enhanced by the addition of 0.1% gelatin into the basal medium as gelatin induces the enzyme capable of degrading PLA (Jarerat et al., 2003). Thermal degradation of PLA can be due to several reasons such as zipper-like depolymerization, intramolecular transesterification, hydrolysis by trace amounts of water, oxidative random main chain cleavage and above 200°C, degradation occur through intra and intermolecular ester exchange, cis elimination, radical, and concerned non-radical reactions (Table 4.3) (Sodergard and Stold, 2002).

4.4 POLY(HYDROXYBUTYRATE) (PHB)

PHB is the supreme common, are gathered as energy and carbon reserves by a large quantity of bacteria. Their biodegradable and biocompatible nature makes these polyesters suitable and easy for many applications. Monomer composition decides the properties of PHAs, so it is great interest recent researches have concluded that, besides PHB a huge assortment of PHAs can be delivered by microbial fermentation PHB is a regular profoundly crystalline thermoplastic though the medium-chain lengths PHAs are elastomers with low softening focuses and a somewhat lower level of crystallinity. An indistinguishable and captivating property of PHAs as for food packaging applications is their low water vapor penetrability, which is near that of LDPE (Figure 4.6).

TABLE 4.3 Commercialized PLA Products

Product	Company Name
Packaging	
Films and trays for biscuits, fruits, vegetables, and meat	Treophan, Natura, IPER, Sainsburys, Suizer, Eco products, RPC
Yogurt cup	Cristallina/Cargill Dow
Trays and bowls for fast food	McDonald's
Rigid transparent packaging of batteries with removable printed film on backside	Panasonic
Envelope with transparent window, paper bag for bread with transparent window	Mitsui, Ecocard
Agriculture and Horticulture	
Mulching films	Novamont, Cargill Dow
Long Life Consumer Good	
Computer keys	Fujitsu
Sapre wheel cover	Toyota
Blanket	Ingeo
Casing of walkman	Sony
Apparel (t-shirt, socks)	FILA/Cargill Dow, KaneboGosen
CD	Sanyo Marvic Media/Lace
Small component of laptop housing	Fujitsu/Lace

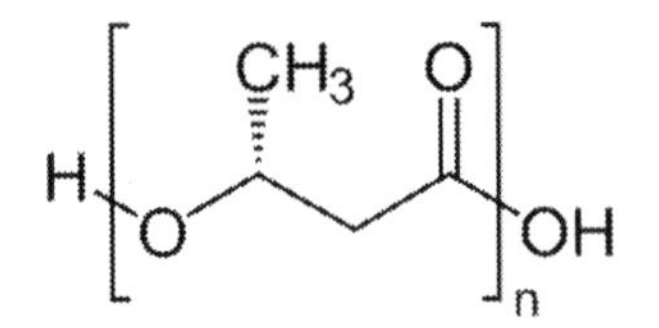

FIGURE 4.6 Structure of poly(hydroxybutyrate) (PHB).

4.4.1 BIOSYNTHESIS OF PHB

Biosynthesis of PHB in bacteria involves three metabolic phases which can be distinguished by:

1. **Phase I:** The process is initiated by the condensation of two molecules of acetyl-CoA to acetoacetyl-CoA, catalyzed by an enzyme 3-ketothiolase (encoded by phbA gene) (Peoples and Sinskey, 1989).
2. **Phase II:** Reduction of acetoacetyl-CoA to 3-hydroxybutyrl-CoA via an enzyme acetoacetyl-CoA reductase (encoded by phbB gene) (Peoples and Sinskey, 1989).

3. **Phase III:** Polymerization of 3-hydroxybutryl-CoA monomer to PHB catalyzed by an enzyme PHB synthase (phbC gene) (Schubert et al., 1988; Slater et al., 1988; Steinbuchel and Schubert, 1989).

4.4.2 PHB PRODUCING BACTERIA

Some *Bacillus* sp., a gram-positive bacterium has the capabilities to synthesis PHB in various culture condition. PHB granules was first observed in the strain of *Bacillus megaterium*, in the year 1926, which has the ability to accumulates PHB under nutrient depletion condition (Katrcolu et al., 2002). PHB was also observed as lipid inclusion constituents in the Bacillus cells (Lemoigne et al., 1950). *Bacillus thuringiensis*, well known for insecticidal endotoxin, also been reported as PHB accumulation (Katırcıoğlu et al., 2003). Several other species such as *Bacillus mycoides* RCJ B-017 (Borah et al., 2002), *Bacillus subtilis*, *Bacillus firmus*, *Bacillus sphaericus* and *Bacillus pumilus* (Katrcolu et al., 2002) have also been observed to accumulate (PHA) in varying amounts (Pal et al., 2009). There are also some aerobic and gram-positive bacteria belonging to the genera *Rhodococcus*, *Nocardia*, and *Streptomyces* has also been reported to synthesis and accumulate PHB under nitrogen restrictive situation (Alvarez et al., 1997). *Ralstonia eutropha, Alcaligenes latus, and* recombinant *Escherichia coli*, gram-negative bacterium have also been involved in the production of PHB (Raetz, 1993). *Streptomyces aureofaciens* has also been employed to accumulate high amount of PHB in the presence of glycerol (Tripathi et al., 2002; Mahishi et al., 2003; Ramachander et al., 2005). Different variety of prokaryotic organisms have been observed to accumulate this polymer which includes *Actinomycetes spp*., Cyanobacteria, heterotrophic, and autotrophic bacteria, photosynthetic anaerobic bacteria (Anderson and Dawes, 1990).

4.4.3 SUBSTRATE FOR PHB PRODUCTION

The commercial production of PHB utilized relatively low cost substrate which includes methanol, beet molasses, ethanol, starch, whey (Kim, 2000; Ghaly et al., 2003; Sharifzadeh et al., 2009), cane molasses as a sole carbon source (Gouda et al., 2001) wheat hydrolysate and fungal extract (Koutinas et al., 2007) or soy cake (Oliveira et al., 2007). Lee and Yu (1997) utilized whey-based medium for the accumulation of a large amount of PHB by examining various recombinant *E. coli* strains. Toledo et al. (1995) used wastewater from olive oil mills as a growth media for *Azotobacter chroococcum* H23 for the production of PHB. Soya and malt squander from lager bottling works plant were additionally used as substrate for the development of *Alcaligens latus* for the synthesis of bioplastics (Yu et al., 1999). Different nitrogen-rich media such as casein hydrolysate, yeast extract, tryptone, corn steep liquor, and collagen hydrolysate utilized in PHB bioconversion via recombinant *E. coli* strains (Table 4.4) (Khanna et al., 2005).

TABLE 4.4 Properties of PHB

Properties	PHB
Molecular weight	500
Glass transition temperature	15°C
Melting temperature	175°C
Crystallinity	80%
Extension to break	6%
Tensile strength	40 MPa
Oxygen permeability	45 cm^3/m^2/atm/day
Resistance to reagent	Weak
Resistance to UV	Good
Solubility	Chloroform and other chlorinated hydrocarbon

4.4.4 BIODEGRADABILITY OF PHB

PHB biodegradability is one of the most deterministic factors to be measured as a biopolymer, which can be determined using various field and laboratory test. By several phenomenon, polymers present in the environment get degraded through hydrolysis, thermal, mechanical, oxidative, and photochemical destruction and biodegradation, whereas PHB biodegradability requirements differs in accordance with their application. The most interesting property of PHB with respect to natural balance mean entirely get degraded into carbon dioxide and water by microorganisms. More than 300 microbial strains were isolated from soil which was capable of degrading PHB, in which denitrifying bacteria plays an important role in degradation (Maergaert et al., 1992). Biodegradability of PHB films have also studied under aerobic, anaerobic, and microaerobic condition in the presence or lack of nitrate by microbial population of soil and sludge from nitrifying/denitrifying reactors, sediment or sludge deposit site, as well as to obtain active denitrifying enrichment culture degrading PHB (Bonartsev et al., 2007). Degradation rate of bioplastic is dependent upon many environmental factors such as pH, temperature, nutrients, and moisture and also on material properties such as crystallinity degree, surface area, and the formulation. PHB biodegradation was also reported under different composting condition, where 18–40% PHB loss was studied after 12 weeks under green manure composting (Harabak, 1992). PHB degradation is also accompanied by the presence of nitric acid, which leads to the formation of butyrate and acetate as intermediate product and carbon dioxide as the end product. Absence of nitric acid also leads to the degradation of PHB, resulting in the formation of butyrate, acetate, and small amount of propionate also.

4.4.5 PHB APPLICATION

PHB has attracted interest for food packaging application due to its certain properties such as renewability, biodegradability, water vapor barrier properties, and oxygen barrier properties. As compared to PLA, PHB act as better light barrier in the visible and ultraviolet light region. The two major drawbacks of PHB, i.e., cost, and narrow melt processing window have limited the use of this biopolymer. It can be overcome by copolymerization with hydroxyvalerate or hydroxyhexanoate. Blending of PHB or PHBV with other biodegradable materials can reduce the brittleness and will produce the small spherulites with improved mechanical properties. Along with the application of packaging of food products, PHB also used in surgical sutures, tissue engineering (Misra et al., 2010), agricultural foils, tablet packaging and microcapsules in therapy due to its biodegradability, eco-friendly, and good barrier properties (Chen, 2011). Hence it act as a good substitute to PP, PE, and also PET in certain application (Kosior et al., 2006).

4.5 SHELF LIFE STUDIES OF FRESH FRUITS AND VEGETABLES (F&V)

Biobased materials will most likely be applied to foods requiring short-term chill storage, such as F&V, since biobased materials present opportunities for producing films with variable CO_2/O_2 selectivity and moisture permeability. Dukalska et al. (2008) studied the capacity time impacts on the quality of fresh strawberries, dark flows, and raspberries. Two strawberry assortments, monetarily developed in Latvia, black current just as late-bearing raspberries, container boxes as control packaging with berries were encased into pockets size of 200×300 mm produced using biodegradable PLA films thickness of 25 (Treofan organization) and 40 μm (MaaG organization) the nuance of oxygen and carbon dioxide in the MAP packaged black currents holder headspace at the storage time it appeared in Table 4.5. As fixing of packages was performed at barometrical ear feel, the underlying substance of CO_2 was assumed near zero and O_2 proportionately as in environment –21%. It was seen the PP plate had the most elevated CO_2 content 18% fixed with OPP film following 25 stockpiling days. Low CO_2 permeability of OPP film might be the reason, which advances CO_2 development in the packaging. It was seen that CO_2 for capacity of berries in the container confines embedded PLA pockets thickness of 40 μm (MaaG organization) was 11 to 12% and O_2–4% which was worthy, which could be assessed as appropriate to equilibrium adapted atmosphere (EMAP) for negligible breathing of natural products at the storage time.

TABLE 4.5 The Dynamics of Oxygen and Carbon Dioxide at the Storage Time in the Headspaces of Different Containers with Packed Black Current

Biodegradable Packaging	Storage Time (days)	CO_2 Content (%)	O_2 Content (%)
PLA pouches thickness of 40 μm (200 × 300 mm)	0	0%	20%
	3	3%	15%
	7	8%	8%
	10	11%	6%
	14	10.25%	5%
	17	10.75%	4.75%
	21	13%	4.5%
	25	12%	0%
Conventional Packaging	0	0%	20%
PP trays sealed with OPP film thickness 40 μm (210 × 148 × 35 mm)	3	10%	10%
	7	12%	7%
	10	13.75%	6%
	14	14%	5%
	17	16%	4.75%
	21	19%	4.75%
	25	18%	4%

Zhou et al. (2016) studied relative quality change of fresh cuts of melons in bio-based and petroleum-based plastic containers during storage. In this investigation, packaging container produced using PLA and PET was tried for their capacity to safeguard the nature of new cut melon under low temperature conditions (10°C). The results showed that the overall quality of fresh-cut melon declined during storage. Melon cubes in both packages showed increases in weight loss, juice leakage, surface color (E) and TA, and decreases in surface color (L*, a*, b*), firmness, SSC, pH, vitamin C and sensory evaluation. No significant differences in color, firmness, pH, TA, or sensory evaluation were found between the different packages at 4°C. However, because of higher water vapor and oxygen permeability of PLA than that of PET, significant differences in color of melon cubes between in the PLA and in PET containers were found on the 7th sampling day of storage at 10°C. The PLA containers preserved the quality of fresh-cut melon better than PET containers at 10°C during 10 d of storage (Table 4.6).

TABLE 4.6 The Water Vapor Transmission Rate (WVTR) and Water Vapor Permeability Coefficient (WVPC) of PLA and PET Containers at 23°C and 46% Relative Humidity Differences (n = 3)

Material	Thickness (μm)	WVTR (g m^{-2} d^{-1})	WVPC (g $m^{-1}d^{-1}Pa^{-1}$)
PLA	270	18.18	3.80×10^{-6}
PET	240	2.09	0.39×10^{-6}

Botondi et al. (2015) studied the effects of packaging consuming two different polymeric trays with hinged lids, polyethylene terephthalate (PET) and polylactic acid (PLA); on fresh-cut and cooked spinach (*Spinacia oleracea*) according to this study, the PLA trays maintained its flavor longer as compared with PET polymeric hinged trays (Table 4.7).

TABLE 4.7 Total Chlorophyll Content of Spinach in PET and PLA Trays

Biodegradable Packaging	Storage Time (days)	Total Chlorophyll (µg/mg f.w*)	mg GAE/100 (g.f.w*)	mg of Spinach /mg of DPPH	Ascorbic Acid (mg of Ascorbic Acid 100 g^{-1} f.w*)
PLA pouches (118 × 172 × 65 mm)					
	0	90	150	0006	21.88
	3	87	220	0008	8.84
	6	80	n.d	n.d	n.d
	8	97	250	0010	5.20
	10	97	n.d	n.d	n.d
	14	85	160	0006	1.46
	16	82	120	0004	0.5
Conventional Packaging	0	90	150	0006	21.88
	3	95	175	0007	3.58
PET 2 polymer used (118 × 172 × 65 mm)	6	85	n.d	n.d	n.d
	8	100	220	0011	4.01
	10	100	n.d	n.d	n.d
	14	90	170	0006	1.40
	16	85	150	0005	0.3

**F.W*: Fresh weight.

4.6 CONCLUSION

The food industry has shown tremendous improvements in the packaging sector since its beginning in the eighteenth century, with the most dynamic and insightful developments occurring in the previous century. These advances have promoted increased nourishment quality and food security. Nowadays, consumers are more conscious about their health, and the modern fast life style have changed the consumer preferences focusing on searching for new insights and innovations in food packaging with more inclination on food safety and perishability of food products. The new innovations are preferably centered towards the deferring oxidation and controlling moisture movement, microbial development, respiration rates, and unstable flavors and fragrances. This concentration parallels that of nourishment packaging appropriation, which has ambitious variation in the key territories of maintainable packaging, utilization of the packaging esteem chain connections for upper hand, and the developing part of sustenance benefit packaging. Biopolymers with complete biodegradability upon disposal have shown significant potential towards usage of biodegradable packaging of fresh F&V. Although new biodegradable packaging material would be beneficial for the consumer from the health perspective but the production cost is still a big threat in its wide commercialization. The evolution of nanocomposites in food packaging is also an innovative tool which may further improve the quality of packaging material with its wide application in F&V processing Industry. The novel approach of adopting biodegradable packaging material in food packaging will be beneficial for the producers, consumers, and environment. The improvement underway innovation of biodegradable packaging material by selecting cheaper raw material, potential strains and downstream processing will further minimize its cost and will make it economical.

KEYWORDS

- **biodegradable**
- **biopolymers**
- **fruits**
- **shelf life**
- **vegetables**

REFERENCES

Alvarez, H. M., Kalscheuer, R., & Steinbüchel, A., (1997). Accumulation of storage lipids in species of *Rhodococcus* and Nocardia and effect of inhibitors and polyethylene glycol. *Fett/Lipid, 99*(7), 239–246. https://doi.org/10.1002/lipi.19970990704.

Anderson, A. J., & Dawes, E. A., (1990). Occurrence, metabolism, metabolic role, and industrial uses of bacterial polyhydroxyalkanoates. *Microbiological Reviews, 54*(4), 450–472. Retrieved from: http://www.ncbi.nlm.nih.gov/pubmed/2087222 (accessed on 19 December 2020).

Auras, R., Harte, B., & Selke, S., (2004). An overview of polylactides as packaging materials. *Macromol. Biosci., 4*, 835–864.

Bonartsev, A. P., Myshkina, V. L., Nikolaeva, D. A., Furina, E. K., & Makhina, T. A., (2007). Biosynthesis, biodegradation, and application of poly (3-hydroxybutyrate) and its copolymers-natural polyesters produced by diazotrophic bacteria. *Communicating Current Research and Educational Topics and Trends in Applied Microbiology*, 295–307. https://doi.org/10.1002/btpr.2247.

Botondi, R., Bartoloni, S., Baccelloni, S., & Mencarelli, F., (2015). Biodegradable PLA (polylactic acid) hinged trays keep quality of fresh-cut and cooked spinach. *Journal of Food Science and Technology, 52*(9), 5938–5945. https://doi.org/10.1007/s13197-014-1695-x.

Bucci, D. Z., Tavares, L. B. B., & Sell, I., (2005). PHB packaging for the storage of food products. *Polymer Testing, 24*(5), 564–571. https://doi.org/10.1016/j.polymertesting.2005.02.008.

Dukalska, L., Muizniece-brasava, S., & Kampuse, S., (2008). Studies of biodegradable polymer material suitability for food packaging applications. *Production*, 64–68.

Gouda, M. K., Swellam, A. E., & Omar, S. H., (2001). Production of PHB by a *Bacillus megaterium* strain using sugarcane molasses and corn steep liquor as sole carbon and nitrogen sources. *Microbiological Research, 156*(3), 201–207. https://doi.org/10.1078/0944-5013-00104.

Hrabak, O., (1992). Industrial production of poly-Î2-hydroxybutyrate. *FEMS Microbiology Letters, 103*(2–4), 251–255. https://doi.org/10.1111/j.1574-6968.1992.tb05845.x.

John, R. P., Gangadharan, D., & Madhavan, N. K., (2008). Genome shuffling of *Lactobacillus delbrueckii* mutant and *Bacillus amyloliquefaciens* through protoplasmic fusion for l-lactic acid production from starchy wastes. *Bioresource Technology, 99*(17), 8008–8015. https://doi.org/10.1016/J.BIORTECH.2008.03.058.

Katırcıoğlu, H., Aslım, B., Yüksekdað, Z. N., Mercan, N., & Beyatlı, Y., (2003). Production of poly-β-hydroxybutyrate (PHB) and differentiation of putative *Bacillus* mutant strains by SDS-PAGE of total cell protein. *African Journal of Biotechnology, 2*(6), 147–149.

Kimura, T., Ishida, Y., Ihara, N., & Saito, Y. N. I. J. J., (2002). Degradability of biodegradable plastic (poly (lactic acid)) products. *Journal of the Japanese Society of Agricultural Machinery, 64*(3), 115–120.

Koutinas, A. A., Xu, Y., Wang, R., & Webb, C., (2007). Polyhydroxybutyrate production from a novel feedstock derived from a wheat-based biorefinery. *Enzyme and Microbial Technology, 40*(5), 1035–1044. https://doi.org/10.1016/J.ENZMICTEC.2006.08.002.

Lee, S., & Yu, J., (1997). Production of biodegradable thermoplastics from municipal sludge by a two-stage bioprocess. *Resources, Conservation and Recycling, 19*(3), 151–164. https://doi.org/10.1016/S0921-3449 (96)01157-3.

López-Rubio, A., Almenar, E., Hernandez-Muñoz, P., Lagarón, J. M., Catalá, R., & Gavara, R., (2004). Overview of active polymer-based packaging technologies for food applications. *Food Reviews International, 20*(4), 357–387. https://doi.org/10.1081/LFRI-200033462.

Mahishi, L. H., Tripathi, G., & Rawal, S. K., (2003). Poly(3-hydroxybutyrate) (PHB) synthesis by recombinant *Escherichia coli* harboring *Streptomyces aureofaciens* PHB biosynthesis genes: Effect of various carbon and nitrogen sources. *Microbiological Research, 158*(1), 19–27. https://doi.org/10.1078/0944-5013-00161.

Oliveira, F. C., Dias, M. L., Castilho, L. R., & Freire, D. M. G., (2007). Characterization of poly (3-hydroxybutyrate) produced by *Cupriavidus necator* in solid-state fermentation. *Bioresource Technology, 98*(3), 633–638. https://doi.org/10.1016/J.BIORTECH.2006.02.022.

Pal, A., Prabhu, A., Kumar, A. A., Rajagopal, B., Dadhe, K., Ponnamma, V., & Shivakumar, S., (2009). Optimization of process parameters for maximum poly (-beta-) hydroxybutyrate (PHB) production by *Bacillus thuringiensis* IAM 12077. *Polish Journal of Microbiology/Polskie Towarzystwo Mikrobiologow = The Polish Society of Microbiologists, 58*(2), 149–154. https://doi.org/10.3906/biy-0808-10.

Pawar, P. A., & Purwar, A. H., (2013). *Biodegradable Polymers in Food Packaging*, (5), 151–164.

Peoples, O. P., & Sinskey, A. J., (1989). Fine structural analysis of the *Zoogloea ramigera* phbA-phbB locus encoding β-ketothiolase and acetoacetyl-CoA reductase: Nucleotide sequence of phbB. *Molecular Microbiology, 3*(3), 349–357.

Popa, M., Mitelut, A., Niculita, P., Geicu, M., Ghidurus, M., & Turtoi, M., (2011). Biodegradable materials for food packaging applications. *Journal of Environmental Protection and Ecology, 12*(4), 1825–1834.

Raetz, C. R., (1993). Bacterial endotoxins: Extraordinary lipids that activate eucaryotic signal transduction. *Journal of Bacteriology, 175*(18), 5745–5753. Retrieved from: http://www.ncbi.nlm.nih.gov/pubmed/8376321 (accessed on 19 December 2020).

Schubert, P., Steinbüchel, A., & Schlegel, H. G., (1988). Cloning of the alcaligenes eutrophus genes for synthesis of poly-beta-hydroxybutyric acid (PHB) and synthesis of PHB in *Escherichia coli*. *Journal of Bacteriology, 170*(12), 5837–5847. https://doi.org/10.1128/JB.170.12.5837-5847.1988.

Slater, S. C., Voige, W. H., & Dennis, D. E., (1988). Cloning and expression in *Escherichia coli* of the *Alcaligenes eutrophus* H16 poly-beta-hydroxybutyrate biosynthetic pathway. *Journal of Bacteriology, 170*(10), 4431–4436. https://doi.org/10.1128/JB.170.10.4431-4436.1988.

Steinbüchel, A., & Schubert, P., (1989). Expression of the *Alcaligenes eutrophus* poly (β-hydroxybutyric acid)-synthetic pathway in *Pseudomonas* sp. *Archives of Microbiology, 153*(1), 101–104.

Tokiwa, Y., & Jarerat, A., (2004). Biodegradation of poly(l-lactide). *Biotechnology Letters, 26*(10), 771–777. https://doi.org/10.1023/B:BILE.0000025927.31028.e3.

Vroman, I., & Tighzert, L., (2009). Biodegradable polymers. *Materials, 2*(2), 307–344. https://doi.org/10.3390/ma2020307.

Yu, P. H. F., Chua, H., Huang, A. L., Lo, W. H., & Ho, K. P., (1999). Transformation of industrial food wastes into polyhydroxyalkanoates. *Water Science and Technology, 40*(1).

Zhou, H., Kawamura, S., Koseki, S., & Kimura, T., (2016). Comparative quality changes of fresh-cut melon in bio-based and petroleum-based plastic containers during storage. *Environment Control in Biology, 54*(2), 93–99. https://doi.org/10.2525/ecb.54.93.

CHAPTER 5

EDIBLE COATING FOR IMPROVEMENT OF HORTICULTURE CROPS

MEENAKSHI GARG,[1] PRABHJOT KAUR,[2] and SUSMITA DEY SADHU[1]

[1]*Bhaskaracharya College of Applied Sciences, Delhi, India, E-mail: meenakshi.garg@bcas.du.ac.in (M. Garg)*

[2]*Shaheed Rajguru College of Applied Sciences, Delhi, India*

ABSTRACT

In the postharvest industry, the preparation of edible coatings with excellent performance and better functionality for the increased shelf life of horticultural produce is a challenging task. Nowadays, consumers demand safe, stable, healthy, and convenient food, and adverse impact of non-biodegradable packaging waste motivated researchers and scientists to work on edible coatings. Edible coatings function by preventing the growth of microorganisms and reducing oxidative and enzymatic changes. Active compounds like essential oils, ascorbic acid inhibit the activity of polyphenol oxidase and thus help to delay enzymatic browning, especially in minimally processed horticultural products. Edible coatings can also be used for encapsulation of bioactive compound and delivery of nutraceutical. Nanoemulsion based edible coatings help in the preparation of a multi-component/ layer system, thereby enhancing the overall benefit of coating. These coatings are environment friendly. This chapter provides information about functions, mechanism of coating, its formulations, applications of nanoemulsions in coating for improvement of quality during postharvest storage of fruits and vegetables (F&V).

5.1 INTRODUCTION

Horticultural produce are perishable food products having active metabolism during postharvest period. About 15% fruits and 50% vegetables crops are lost after harvest on global scale (FAO, 2011), due to spoilage by microorganisms. In developing countries like Pakistan, India, etc., this percentage is increased because correct technologies are not available for storage. Deterioration of quality and postharvest diseases increases after harvest and these changes favor the growth of pathogens, which causes losses through the supply chain. Hence it is very important to extend the shelf life of horticulture produce by retarding physical, pathological, and physiological deteriorative processes (Robertson, 2013). The development of decay control measures to improve and maintain the quality of fruits and vegetables (F&V) and to protect them against diseases at the retail level will be beneficial to reduce post harvest losses.

Edible packaging plays a crucial role in both the preservation and handling approaches to maximizing the shelf life of horticulture produce. Edible packaging material consists of edible sheets, coatings, films, pouches, etc. Edible films or sheets are stand-alone structures having thickness <254 μm for films and for sheets, thickness should be >254 μm. These might be sealed and edible pouches can be formed or can be used in-between food components. Edible coatings are usually formed on the surface of food directly in the form of thin layers (Janjarasskul and Krochta, 2010). Edible coatings can be applied on a variety of products to control the oxidation process, gas exchange, and moisture transfer (Dhall, 2013). An ideal edible coating is defined as that which can extend and improve the shelf life of fresh horticultural produce without affecting their quality.

5.2 FUNCTIONS OF EDIBLE COATINGS

- These are environmentally friendly technology and can provide an additional protective coating, which can give the same effect as modified atmospheric storage by reducing quality and quantity losses.
- Several active ingredients such as flavors, nutrients, anti-browning agents, spices, etc., can be added into the matrix of the polymer of edible coating and can be consumed with food and can reduce the pathogenic growth on food.
- Nutritional and sensory parameters can be improved by edible coatings (Dhall, 2013).

- It can improve the shelf life of fruit and vegetable by reducing moisture content and migration of solutes, respiration, gas exchange, oxidative changes and it also suppresses physiological disorders.
- Edible coatings are composed of biodegradable raw material. Hence reduces synthetic packaging waste.

5.3 ADVANTAGES AND DISADVANTAGES OF EDIBLE COATING

5.3.1 ADVANTAGES OF EDIBLE COATING

- Retention of color, flavor, acid, and sugar are improved by edible coating in F&V.
- It reduces firmness loss and weight loss.
- Quality of fruit and vegetables are maintained during storage.
- Edible coatings contain many types of nutrients, antioxidants, pigments, etc., that give health benefits if consumed with F&V.
- It reduces packaging waste.
- It can be used in multilayer material with inedible films also in which edible coating would be the inner layer which would be the contacting layer with food.
- Edible coating can be employed for packaging of food small portions that presently are not packaged in units such as nuts, peas, beans, and strawberries (Robertson, 2013; Kaur et al., 2016).

5.3.2 DISADVANTAGES OF EDIBLE COATING

- Many edible coatings are usually hygroscopic, which increases microbial growth if not monitored properly.
- Normal ripening process is disturbed by edible coatings due to barrier properties against gases which causes anaerobic respiration.
- Thick edible coatings causes off flavor because of oxygen prohibition (Raghav et al., 2016).

5.4 MECHANISM OF EDIBLE COATING

Respiration in F&V is present even after harvesting, used oxygen is not replaced quickly by edible coating. It produces carbon dioxide, and this will accumulate in the horticulture produce and cannot escape because of edible coating and produces anaerobic respiration that requires only 1–3% oxygen. Ethylene production is disrupted and causes delay in ripening. Physiological loss of water is also minimized, and the produce remains fresh for a longer time. Several studies showed that edible coating forms a thin semi-permeable film on the surface of fruit and vegetable and delay the rate of respiration, decreases weight loss and maintain the overall quality and prolong the shelf life (Romanazzi et al., 2017).

5.5 REQUIREMENTS FOR EDIBLE COATING

Edible coatings cannot be applied always on the surface of fruit or vegetables. It can be used on fresh-cut or whole fruit or vegetables (Table 5.1).

TABLE 5.1 Examples of Fruits and Vegetables for Edible Coating

Whole Fruits	Whole Vegetables	Fresh-Cut Fruits	Fresh-Cut Vegetables
Apple passion fruit, avocado, kinnow, grapefruit, peach, lemon, lime, etc.	Bell pepper, melons, cucumber, tomato, etc.	Fresh-cut pear, fresh-cut peach, Fresh-cut apple	Fresh-cut potato, fresh-cut tomato slices, minimally processed Carrot, fresh-cut green lettuce, minimally processed onion, Fresh-cut muskmelon and cantaloupe, cabbage, etc.

The edible coating must be water-resistant and should remain intact and cover horticulture product properly after application. Oxygen level should be in between 1 and 3% to avoid shifting from aerobic to anaerobic respiration. It could improve color, appearance, reduce water vapor permeability, and maintain structural integrity. Its melting point should be above 40°C without decomposition. It should be nonsticky, easily emulsifiable, and have good drying performance. It should be economical, low viscous and translucent to opaque, can tolerate little pressure also (Dhall, 2013).

5.6 APPLICATION OF EDIBLE COATING

Coatings are formed from substances which have film-forming properties. Substances are dissolved in water, alcohol, or in combinations throughout the manufacturing process. Antimicrobial agents, colorants, plasticizers, flavors, etc., are fused together. Effective coating can be obtained first by synthesizing and then by drying the solution at a specific temperature and RH. Coating is usually applied by spraying, brushing, dipping, extrusion, and by solvent casting. Material for coating are usually derived from eco friendly and sustainable material (Hassan et al., 2018). The dipping method is commonly used for applying edible coatings on horticultural produce, by dipping for 5–30 seconds in coating solution and gives excellent results on fruits (Valverde, 2005). Brushing method gives good result on beans, strawberry, and berries. Spray coating includes a spraying nozzle and batch tank to disperse coating on fruit and vegetables when they pass through a conveyor roller (Debeaufort and Voilley, 2009).

5.7 COMPONENTS AND TYPES OF EDIBLE COATING

Edible coatings are synthesized from polysaccharides, proteins, and lipids. They act as a barrier against moisture and oxygen during handling and storage. It increases safety because of bioactive compounds present in it.

1. **Polysaccharide Based:** These are naturally occurring polymers coatings includes cellulose, starch, pullulan, chitosan, and pectin and its derivatives. Polysaccharides give hardness, compactness, crispiness, adhesiveness, and thickening quality to of edible coatings. These are synthesized from polymer chain, having good barrier properties (Table 5.2).

TABLE 5.2 Components and Types of Edible Coating

Types of Edible Coating			
Polysaccharide Based	**Lipid-Based**	**Protein-Based**	**Composite**
• Cellulose and its derivative-based • Chitin and chitosan based • Starch-based • Pectin based • Pullulan based • Alginate based	• Waxes and paraffin-based • Acetoglyceride based • Shellac resin based	• Corn zein based • Gelatin based • Wheat gluten-based • Whey protein-based	• Bilayers • Conglomerate

Source: Raghav et al. (2016); with permission.

2. **Protein-Based Edible Coating:** These coatings provide good barrier properties against oxygen at low humidity. It is hydrophilic in nature. It has good mechanical and organoleptic properties. Plant-based material includes zein, gluten, soy protein, etc., and animal-based proteins are albumin, collagen, etc. Casein based edible coatings are popular for fruit and vegetables because of their high nutritional quality, good sensory properties, and good protection against their surrounding environment. Whey proteins with the addition of plasticizer create flexible and transparent water-based edible coatings having excellent aroma, and fats and oil barrier properties at low relative humidity (RH). Whey protein coatings because of their hydrophilic nature are not so effective against moisture (Vargas et al., 2008).
3. **Lipid-Based Edible Coating:** Most common type of lipid-based coatings include cocoa-based material, acetylated glycerides, fatty acid and alcohols, lacs, and waxes. These types of coatings have a preservative effect and provide a glossy and shiny appearance. Generally, paraffin wax, vegetable or mineral oil, bee wax are used for lipid coating. These coatings provide good moisture barrier properties but are not easily applied on F&V because of its greasy nature that gives rancid flavor. Polysaccharides, lipid, and protein are used in combination to improve the barrier properties (Robertson, 2009).
4. **Composites Based Edible Coating:** These are multi-component coatings contain a combination of proteins, polysaccharides, and lipids and are used to improve mechanical strength and other barrier properties of coatings (Phan et al., 2008; Robertson, 2009). Either different or same type of coatings is combined in bilayer edible coating. These types of coatings generally include propylene, protein, water waxes, etc. The aim of composite coating is to improve the mechanical and permeability properties according to the requirement of the industry. These films are applied in the suspension form, emulsion or dispersion in multilayer coating (Table 5.3) (Dhall, 2013).

5.8 FUNCTIONAL PROPERTIES OF EDIBLE COATINGS

Edible coatings are biodegradable and edible. Main functional properties are permeation, migration, moisture, and air barrier functions and both physical and mechanical protection. These properties increase their use for food quality preservation and for extension of shelf life. Edible coatings are carriers of active compounds that should be released in a controlled and systemic way to promote antimicrobial activity and can be used for spoilage control and for safety enhancement. The most important characteristic of edible coating is edibility and biodegradability, and it can be achieved by the use of food grade ingredients. Using edible coating, F&V can be protected by bruising, tissue damage, or any physical injury caused by pressure, impact, vibration, or any other mechanical factors. Edible coating maintains the integrity of food (Palou et al., 2015).

TABLE 5.3 Type and Composition of Edible Coating

Types of Edible Coating	Composition	Application	References
Polysaccharides	Cassava starch	Strawberry coating	Garcia et al. (2012)
	Chitosan	Asparagus, Pomegranate	Qiu et al. (2012)
	Pectin	Melon, mango	Ferrari et al. (2013)
	Alginate	Cherry	Diaz-Mula et al. (2012)
Proteins	Soy protein isolates	Fresh cut eggplant coating	Ghidelli et al. (2014)
	Zein from corn	Fresh fruits	Bai et al. (2003)
	Gluten from wheat	Fruits	Chao et al. (2007)
	Whey proteins	Fruits	Mei and Zhao (2003)
Lipids	Shellac	Fresh fruits and vegetables	Bai et al. (2003)
Composite	Pectin, cellulose, monoglyceride	Fresh cut apple	Wong et al. (1994)

5.9 APPLICATION OF EDIBLE COATING ON HORTICULTURE CROPS

Edible coatings are gaining importance to reduce the deteriorative changes, especially in the case of minimally processes fruits and vegetable. The minimal processing cause's tissue damage as a result, rate of enzymatic and oxidative reaction increase. High water activity is another important factor in determining the shelf life of minimally processed F&V (Galus and Kadzińska, 2015). Horticulture crop is also susceptible to microbial growth and proliferation during processing and post-processing operations especially storage (Sanchís et al., 2016). The extension of shelf life is attributed due to the ability of edible coatings in tumbling moisture relocation, gas exchange, reducing oxidative and enzymatic changes and physiological disorders (Robles-Sánchez et al., 2013).

Conventionally used synthetic wax coatings can be replaced by suitable edible coating, as these get lost during washing and handling. Oranges which were wax coated naturally were compared with ones coated with gelatin, Persian gum and shellac coated oranges. With increased storage time weight loss, reduction in fruit firmness, pH, total soluble solids, total phenolic content and total antioxidant activity were greater in wax-coated oranges as compared to edible coatings and best results with shellac coating indicating it can be commercialized as a replacement to synthetic wax (Khorram et al., 2017).

Edible films and coatings have been widely applied to extend the shelf life of horticultural crops. They have been successfully applied to extend the shelf life of F&V and moreover in preserving the quality of foods (Cazón et al., 2017). Diverse coatings (lipids, protein, polysaccharide, essential oils, etc.), have been reported in literature to delay Physico-chemical changes associated with horticulture crops. Aloe vera gel containing coatings have been reported to extend shelf life of various horticulture crops like pomegranate arils (Martínez-Romero et al., 2013), sweet cherry (Martínez-Romero et al., 2006), blueberry (Vieira et al., 2016). In a study conducted on table grapes, (Valverde et al., 2005), the storability of table grapes could be extended up to 35 days (at 1°C) in comparison to uncoated clusters which could be kept for 7 days at 1°C. Edible coatings especially alginate, cellulose, soy protein isolate, whey protein, albumin, etc., have been reported to reduce oil absorption/incorporation during frying (Falguera et al., 2011).

5.10 PROPERTIES OF EDIBLE COATINGS FOR QUALITY IMPROVEMENT

5.10.1 ANTIMICROBIAL

One of the main applications of edible coatings in horticulture crops is to prolong shelf life by diminishing the losses. Fresh F&V are susceptible to microbial spoilage. A lot of edible coatings with antimicrobial agents is used to retard microbial growth. Pullulan is a natural microbial polysaccharide and pullulan films are tasteless, transparent, and elastic with good barrier properties making them an ideal candidate for edible coating. Pullulan film containing sweet basil extract showed low antibacterial activity against mesophilic bacteria and good inhibition against *Rhizopus arrhizus* on apple. Sweet basil extract relocates the polyphenols to the inner microbial cell, thereby disturbing the nucleic acid synthesis (Synowiec et al., 2014). Natural antimicrobials like malic oil, essential oils (from cinnamon, palmarosa, and lemongrass) were incorporated into alginate-based coating applied on melon. Malic acid proved to be effective improvising shelf life of melon on microbiological and physic-chemical parameters. Essential oils were efficacious in inhibiting natural microflora and *Salmonella enteritidis* (Raybaudi-Massilia et al., 2008). Aloe vera based coatings are gaining popularity over traditional edible coatings, as it is antimicrobial, safe, and easily prepared. Chitosan is another natural biopolymer which antifungal and antimicrobial. Chitosan nanoparticles coated on brinjal, tomato, and chilly showed decrease in weight loss and antimicrobial activity against plant pathogens like *Rhizoctonia solani, Fusarium oxysporum, Colletotrichum acutatum,* and *Phytophthora infestans* (Raybaudi-Massilia, Mosqueda-Melgar, and Martín-Belloso, 2008).

5.10.2 ANTI-BROWNING

Many fresh fruits and vegetable fetch lower price in the market because of color. Horticulture crops undergo surface browning, which is an indicator of inferior quality and the main factor limiting the quality of fresh cut fruits and vegetables. Polyphenol oxidase causes oxidation of phenols to quinones in the presence of oxygen, which causes color discoloration, also called enzymatic browning. Main indicators of browning are polyphenol oxidase activity, whiteness index, and total phenols. High total phenol content implies a lower rate of browning in horticulture crop. Xing et al., 2010) in their study on fresh-cut lotus root slices reported prevention of browning by chitosan-based coating under MAP storage for a period of 8 days. Fresh cut lotus root slices undergo discoloration/ browning due to surface-induced oxidative and enzymatic reactions. Ascorbic acid (2%) and citric acid (1%) were added as antibrowning agents in chitosan coating. The L* value for lotus root slices tends to decrease with storage period, but the decrease was far less in coated samples kept under MAP storage. Antibrowning agents in coating led to inhibition of Polyphenol oxidase activity.

In another study on fresh-cut 'Flor de Invierno' pears (Mart, 2008), alginate, gellan, and pectin based coating incorporating antioxidants (N-acetylcysteine and glutathione) showed reduced ethylene production, lower microbial count, increased vitamin C and total phenolic content in comparison to control samples for a period of 14 days at 4°C. The common additives used as anti-browning agents are listed in Table 5.4. Edible composite coating from whey protein isolate, whey protein concentrate, and beeswax showed antibrowning effect on apple slices (Perez-Gago et al., 2005).

TABLE 5.4 Anti-Browning Effect of Different Types of Coating on Fruits and Vegetables

Coating/ Technology	Active Component	Example of Fruit/ Vegetable	References
Chitosan	Citric acid and ascorbic acid	Fresh-cut lotus root slices	Xing et al. (2010)
Konjac glucomannan	Pineapple fruit extract	Fresh-cut rose apple	Supapvanich et al., (2012)
Alginate, pectin, and gellan	N-acetylcysteine and glutathione	Fresh-cut pears	Mart (2008)
Alginate and gellan	N-acetylcysteine and glutathione	Fresh-cut Fuji apples	Tapia et al. (2007)
Xanthan gum	Cinnamic acid	Fresh-cut pears	Sharma and Rao (2015)
Alginate	Ascorbic acid and citric acid	Fresh-cut mangoes	Robles-Sánchez et al. (2013)
Apple pectin	Citric acid and calcium chloride	Fresh-cut persimmon	Sanchís et al. (2016)
Carboxymethyl cellulose	Calcium chloride and ascorbic acid	Fresh-cut apples	Koushesh and Sogvar (2016)

5.10.3 NUTRITIONAL QUALITY

Edible coating limits the growth of microorganism, oxidative, and enzymatic reactions, so that the nutrition quality of indigenous horticultural crop can be sustained. Often natural antioxidants are incorporated in edible coatings that widen their functionality. Edible coatings with active compound as ascorbic acid have significantly higher levels of total phenol. Ascorbic acid also helps in reduction of polyphenol oxidase activity by reduction of quinone formation from phenols and reduction of *o*-benzoquinones back to *o*-diphenols. Similarly, rise in flavonoid level in ascorbic acid dip was reported (Moreíra et al., 2015). In another study, Chinese winter jujube was preserved with alginate based edible coating incorporated with tea polyphenols. The study concluded preservation of total chlorophyll, ascorbic acid as well as total phenolic compounds (Zhang et al., 2016). Strawberries were studied for the effect of edible coating (limonene and curcumin) on total phenolic count. Strawberries coated with limonene had higher total phenolic compound initially, but a significant increase was observed on storage in curcumin-coated strawberries (Dhital et al., 2017).

5.10.4 IMPROVISING TEXTURE/AS FIRMING AGENT

Edible coatings have a huge effect on the texture, appearance, color, and firmness of horticultural crop. Loss of firmness indicates deterioration of fresh produce due to hydrolysis of pectic acids in the walls of the cell beside consequent weight loss because of fluid loss. The consequence of edible coating frequently tips to intensification in firmness. Fresh cut papaya pieces coated with gellan and ascorbic acid showed double firmness in comparison to uncoated pieces (Tapia et al., 2008). Carboxymethyl cellulose coating when collectively used alongside calcium chloride with ascorbic acid on thin slices of apple held firmness of flesh. Calcium chloride functions by reacting with pectin of tissue to from calcium pectate, which imparts strength to the cell wall (Koushesh and Sogvar, 2016). Results on the same line have been reported in blueberries (Mannozzi et al., 2017). Combination of alginate with pectin resulted in an edible covering that improvised and maintained the firmness up to a period of 10 storage days. Carnauba wax with sodium alginate coating on eggplant showed minimum decrease in firmness in comparison to the samples without coating for a storage period of 12 days (Singh et al., 2016).

However, all positive results are not reported in the literature. Raspberries coated with pectin and alginate with calcium chloride did not report retention of firmness to a very high extent (Guerreiro et al., 2015a).

5.11 INNOVATIONS IN EDIBLE COATING FOR FRESH FRUITS AND VEGETABLE

Aloe vera based coatings are becoming popular since they are safe and environment friendly. Studies report various species of aloe vera have antimicrobial activity, especially against fungi. The antifungal activity is associated with the content of major phenolic compound, aloin, found in aloe leaves (Zapata et al., 2013). Guillén et al. (2013) reviewed the efficiency of aloe vera coating on peaches and plums. Coatings were helpful in delaying ethylene production, weight loss reduction, and acidity. A significant 50–70% ethylene inhibition was observed over the control samples. Benítez et al. (2015), compared aloe vera coatings with traditional coating (chitosan and sodium alginate) on kiwifruit slices. It was concluded that aloe vera coatings were fruitful in extending the shelf life as well as prohibited the loss of ascorbic acid and maintained the sensory attributes throughout the storage period. Similar effect was observed with pomegranate arils. Aloe vera gel in combination with citric acid and ascorbic acid locating under MAP storage led to lower CO_2 concentration, lower reduction in firmness and significantly lesser count of mesophilic aerobes, yeasts, and molds (Table 5.5) (Martínez-Romero et al., 2013).

TABLE 5.5 Advancement in Aloe Vera based Edible Coatings

Coating	Example of Fruit/Vegetable	References
Chitosan	Blueberries	Vieira et al. (2016)
Aloe vera with citric acid and ascorbic acid	Pomegranate arils	Martínez-Romero et al. (2013)
Aloe vera and aloe arborescence gels	Peaches and plums	Guillén et al. (2013)
Aloe vera	Fresh-cut kiwifruit slices	Benítez et al. (2015)
Aloe vera	Eggplant	Amanullah et al. (2016)
Aloe vera gel	Raspberry	Hassanpour (2015)
Aloe vera gel	Tomatoes	Ortega-Toro et al. (2017)
Aloe vera gel	Apple slices	Chauhan et al. (2011)
Aloe vera gel	Hayward kiwifruit	Benítez et al. (2013)
Aloe vera gel	Sweet cherry	Martínez-Romero et al. (2006)

Edible coating can be used in combination with other processing technologies. In one such study, commercial edible coating (Sta-fresh 2505™) was used in combination with gamma irradiation on golden-yellow and purple-red tamarillo (*Solanum betaceum* Cav.) fruits. The combination had a synergistic effect with weight loss reduced by 48%, firmness, and appearance higher by 70% and 40% in comparison to control samples (Abad et al., 2017). Pulsed light treatment in combination with gellan gum enriched with apple fiber-based coating retarded microbial growth and maintained sensory quality of 'Golden Delicious' apples. There was a significant reduction in softening and browning of apple pieces through storage (Moreíra et al., 2015).

5.12 ESSENTIAL OIL AND NANOEMULSION BASED EDIBLE COATING

Essential oils or their compounds are added in edible coatings to enhance the functionality of edible coating. Essential oils have proven inhibitory effect on microbial thereby; higher shelf life can be achieved. Citral and eugenol have been successfully incorporated into alginate and pectin based coating for raspberries (Guerreiro et al., 2015a). Citral and eugenol incorporated alginate coating showed similar effect in firmness of *Arbutus* beery fruit on first 14 days of storage (Guerreiro et al., 2015b). Thereby extending the shelf life of horticultural crop.

Nanoemulsion provide an excellent vehicle for delivery of lipophilic active ingredients in an aqueous medium with unique physicochemical and functional virtues. Nanoemulsion can be designed for specific food and may enhance the bioavailability of active compound. Studies show nanoemulsion based coating is more effective than conventional coatings. Nanoemulsion opens avenues for incorporation of varied compounds and further application in the food industry (Acevedo-Fani et al., 2017). In a recent study conducted on fresh-cut Fuji apples, lemon grass essential oil containing nanoemulsion coating exhibited faster and greater inactivation of *Escherichia coli* during storage, reduced respiration and ethylene production rates (Salvia-Trujillo et al., 2015). Nanoemulsion of thyme, lemongrass oil dispersed in sodium alginate solution was used to prepare transparent, high water vapor resistance and better flexibility than their individual counterparts. The film thus formed had the strongest antimicrobial effect against *E. coli* up to 4.71 log reductions after 12 hours (Acevedo-Fani et al., 2015).

5.13 CHALLENGES AND OPPORTUNITIES IN EDIBLE COATING FOR IMPROVEMENT OF HORTICULTURE CROP

Edible coatings have not been utilized as a vehicle. There is a lot of untapped potential for the same, especially for minimally processed F&V. Little research has been done on employing edible coatings to enhance functional properties like as a carrier for nutraceutical, firming agent and as a carrier for color, preservatives, and volatile compounds. More studies are needed in the field of edible coating to understand how the modifications can be made to extend the shelf life of F&V. New formulations with incorporation of active/ functional compounds should be worked on. Detailed studies for changes in bioactivity of compounds and antioxidant activity on edible coating of various F&V are needed (Robles-Sánchez et al., 2013). Lot of studies reported in literature lack effect of edible coatings on the sensorial quality of food. Thus, detailed studies for the same are needed.

Functional parameters of horticulture crops can be improvised with the addition of active ingredients rather than using coating alone in a system (Saba and Sogvar, 2016). The choice of incorporation of active ingredients is limited as only generally regarded as safe and additives that meet international regulations, can be added in edible coatings (Sanchís et al., 2016). More research is needed to study the influence of edible coatings on bioactive compounds of horticulture produce (Mannozzi et al., 2017). Research for active compounds from non-conventional sources for incorporation in edible coating is need of the hour.

Emulsion and nanoemulsion based coatings have gained popularity in recent times. Nanoemulsion is an upcoming field with varied challenges to address. The usage of nanoemulsion is scarce in the food industry. The biggest challenge is the usage of nanoemulsion on solid food surface (Salvia-Trujillo et al., 2015). These films/coating show better mechanical property in comparison to pure counterparts. Few lipid-based coatings affect the sensorial quality of horticultural crop by imparting a waxy sensation (Galus and Kadzińska, 2015). Edible coatings, especially for minimally processed F&V provide shelf stability up to 12–15 days on average. The coatings should be developed and worked upon so as to give stability and also lessen any undesirable modifications for longer durations.

In addition, edible coatings have limited usage in the food industry because of their lesser physical properties like mechanical properties and gas barrier properties with respect to synthetic polymers. Thus, the biggest challenge is to bring the barrier and mechanical properties equivalent to synthetic polymers.

5.14 CONCLUSION

An edible coating is an upcoming arena gaining popularity among researchers and common masses. They can be successfully employed to extend shelf life and storage stability of fresh horticultural produce and minimally processed produce in a natural and healthy way. Edible coatings function by inhibiting microbial growth, reducing oxidative and enzymatic changes. Active compounds like essential oils, ascorbic acid inhibit polyphenol oxidase activity and thus delaying browning, especially in the case of minimally processed F&V. Edible coatings can also be used for encapsulation of bioactive compounds and delivery of nutraceutical. Nanoemulsion based edible coatings helps in the preparation of a multi-component/layer system, thereby enhancing the overall benefit of the coating. Further, research is needed to articulate nanoemulsion that can be used on solid surfaces as well as for newer active compounds, including that from non-conventional sources.

KEYWORDS

- **antimicrobial**
- **chitosan**
- **edible coating**
- **nanoemulsion**
- **nutraceuticals**
- **postharvest**

REFERENCES

Abad, J., Valencia-Chamorro, S., Castro, A., & Vasco, C., (2017). Studying the effect of combining two non-conventional treatments, gamma irradiation, and the application of an edible coating, on the postharvest quality of tamarillo (*Solanum betaceum* Cav.) fruits. *Food Control, 72*, 319–323. Elsevier Ltd. doi: 10.1016/j.foodcont.2016.05.024.

Acevedo-Fani, A., Salvia-Trujillo, L., Rojas-Graü, M. A., & Martín-Belloso, O., (2015). Edible films from essential-oil-loaded nanoemulsions: Physicochemical characterization and antimicrobial properties. *Food Hydrocolloids, 47*, 168–177. Elsevier Ltd. doi: 10.1016/j.foodhyd.2015.01.032.

Acevedo-Fani, A., Soliva-Fortuny, R., & Martín-Belloso, O., (2017). Nanoemulsions as edible coatings. *Current Opinion in Food Science, 15*, 43–49. doi: 10.1016/j.cofs.2017.06.002.

Amanullah, S., Jahangir, M. M., Ikram, R. M., Sajid, M., Abbas, F., & Mallano, A. I., (2016). Aloe vera coating efficiency on shelf life of eggplants at differential storage temperatures. *Journal of Northeast Agricultural University (English Edition), 23*(4), 15–25. doi: https://doi.org/10.1016/S1006-8104(17)30003-X.

Bai, J. H., Alleyne, V., Hagenmaier, R. D., Mattheis, J. P., & Baldwin, E. A., (2003). Formulation of zein coatings for apples (*Malus domestica* Borkh). *Postharvest Biol. Tec., 28*, 259–268.

Bai, J. H., Hagenmaier, R. D., & Baldwin, E. A., (2003b). Coating selection for 'delicious' and other apples. *Postharvest Biol. Tec., 28*(3), 381–390.

Benítez, S., Achaerandio, I., Pujolà, M., & Sepulcre, F., (2015). Aloe vera as an alternative to traditional edible coatings used in fresh-cut fruits: A case of study with kiwifruit slices. *LWT-Food Science and Technology, 61*(1), 184–193. doi: 10.1016/j.lwt.2014.11.036.

Benítez, S., Achaerandio, I., Sepulcre, F., & Pujolà, M., (2013). Aloe vera based edible coatings improve the quality of minimally processed "Hayward" kiwifruit. *Postharvest Biology and Technology, 81*, 29–36. doi: https://doi.org/10.1016/j.postharvbio.2013.02.009.

Cazón, P., Velazquez, G., Ramírez, J. A., & Vázquez, M., (2017). Polysaccharide-based films and coatings for food packaging: A review. *Food Hydrocolloids, 68*, 136–148. doi: 10.1016/j.foodhyd.2016.09.009.

Chauhan, O. P., Raju, P. S., Singh, A., & Bawa, A. S., (2011). Shellac and aloe-gel-based surface coatings for maintaining keeping quality of apple slices. *Food Chemistry, 126*(3), 961–966. doi: https://doi.org/10.1016/j.foodchem.2010.11.095.

Cho, S. Y., Park, J. W., Batt, H. P., & Thomas, R. L., (2007). Edible films made from membrane processed soy protein concentrates. *LWT-Food Sci. Technol., 40*, 418–423.

Debeaufort, F., & Voilley, A., (2009). Lipid-based edible films and coatings. In: Embuscado, M. E., & Huber, K. C., (eds.), *Edible Films and Coatings for Food Applications* (pp. 135–168). New York: Springer.

Dhall, R. K., (2013). Advances in edible coatings for fresh fruits and vegetables: A review. *Critical Reviews in Food Science and Nutrition, 53*(5), 435–450. doi: 10.1080/10408398.2010.541568.

Dhital, R., Joshi, P., Becerra-Mora, N., Umagiliyage, A., Chai, T., Kohli, P., & Choudhary, R., (2017). Integrity of edible nano-coatings and its effects on quality of strawberries subjected to simulated in-transit vibrations. *LWT-Food Science and Technology, 80,* 257–264. Elsevier Ltd. doi: 10.1016/j.lwt.2017.02.033.

Diaz-Mula, H. M., Serrano, M., & Valero, D., (2012). Alginate coatings preserve fruit quality and bioactive compounds during storage of sweet cherry fruit. *Food and Bioprocess Technology, 5*(8), 2990–2997.

Falguera, V., Quintero, J. P., Jiménez, A., Muñoz, J. A., & Ibarz, A., (2011). Edible films and coatings: Structures, active functions and trends in their use. *Trends in Food Science and Technology, 22*(6), 292–303. doi: 10.1016/j.tifs.2011.02.004.

FAO, (2011). Global food losses and food waste. In: *International Congress "Save Food."* D€usseldorf, Germany.

Ferrari, C. C., Sarantopoulos, C. I. G. L., Car-mello-Guerreiro, S. M., & Hubinger, M. D., (2013). Effect of osmotic dehydration and pectin edible coatings on quality and shelf life of fresh-cut melon. *Food Bioprocess Technology, 6*(1), 80–91.

Galus, S., & Kadzińska, J., (2015). Food applications of emulsion-based edible films and coatings. *Trends in Food Science and Technology, 45*(2), 273–283. doi: 10.1016/j.tifs.2015.07.011.

Garcia, L. C., Pereira, L. M., Saran-Topoulos, C. I. G. D. L., & Hubinger, M. D., (2012). Effect of antimicrobial starch edible coating on shelf life of fresh strawberries. *Packaging Technology and Science, 25*(7), 413–425.

Ghidelli, C., Mateos, M., Rojas-Argudo, C., & Pérez-Gago, M. B., (2014). Extending the shelf life of fresh-cut eggplant with a soy protein-cysteine based edible coating and modified atmosphere packaging. *Postharvest Biol. Tec., 95*, 81–87.

Gianfranco, R., Erica, F., Silvia, B. B., & Dharini, S., (2017). Shelf-life extension of fresh fruit and vegetables by chitosan treatment. *Critical Reviews in Food Science and Nutrition, 57*(3), 579–601. http://dx.doi.org/10.1080/10408398.2014.900474.

Guerreiro, A. C., Gago, C. M. L., Faleiro, M. L., Miguel, M. G. C., & Antunes, M. D. C., (2015a). Raspberry fresh fruit quality as affected by pectin- and alginate-based edible coatings enriched with essential oils. *Scientia Horticulturae, 194*, 138–146. Elsevier B. V. doi: 10.1016/j.scienta.2015.08.004.

Guerreiro, A. C., Gago, C. M. L., Faleiro, M. L., Miguel, M. G. C., & Antunes, M. D. C., (2015b). The effect of alginate-based edible coatings enriched with essential oils constituents on *Arbutus unedo* L. fresh fruit storage. *Postharvest Biology and Technology, 100*, 226–233. Elsevier India Pvt. Ltd. doi: 10.1016/j.postharvbio.2014.09.002.

Guillén, F., Díaz-Mula, H. M., Zapata, P. J., Valero, D., Serrano, M., Castillo, S., & Martínez-Romero, D., (2013). Aloe arborescens and Aloe vera gels as coatings in delaying postharvest ripening in peach and plum fruit. *Postharvest Biology and Technology, 83*, 54–57. Elsevier B.V. doi: 10.1016/j.postharvbio.2013.03.011.

Hassan, B., Ali, S. C. S., Abdullah, I. H., Khalid, M. Z., & Naseem, A., (2018). Recent advances on polysaccharides, lipids, and protein-based edible films and coatings: A review. *International Journal of Biological Macromolecules, 109*, 1095–1107.

Hassanpour, H., (2015). Effect of aloe vera gel coating on antioxidant capacity, antioxidant enzyme activities, and decay in raspberry fruit. *LWT-Food Science and Technology*, *60*(1), 495–501. doi: https://doi.org/10.1016/j.lwt.2014.07.049.

Janjarasskul, T., & Krochta, J. M., (2010). Edible packaging materials. *Annu. Rev. Food Sci. Technol., 1*, 415–448. doi: 1146/annurev.food.080708.100836.

Kaur, P., Garg, M., Aditi, & Sadhu, S. D., (2016). Advancement in conventional packaging-edible packaging. *World Journal of Pharmaceutical and Life Sciences, 2*(3), 160–170.

Khorram, F., Ramezanian, A., & Hosseini, S. M. H., (2017). Shellac, gelatin, and Persian gum as alternative coating for orange fruit. *Scientia Horticulturae, 225*, 22–28 Elsevier. doi: 10.1016/j.scienta.2017.06.045.

Koushesh, S. M., & Sogvar, O. B., (2016). Combination of carboxymethyl cellulose-based coatings with calcium and ascorbic acid impacts in browning and quality of fresh-cut apples. *LWT-Food Science and Technology, 66*, 165–171. Elsevier Ltd. doi: 10.1016/j.lwt.2015.10.022.

Lluís, P. L., Silvia, A., Chamorro, V., María, B., & Gago, P., (2015). Antifungal edible coatings for fresh citrus fruit: A review. *Coatings*, *5*, 962–986. doi: 10.3390/coatings5040962.

Mannozzi, C., Cecchini, J. P., Tylewicz, U., Siroli, L., Patrignani, F., Lanciotti, R., Rocculi, P. et al., (2017). Study on the efficacy of edible coatings on quality of blueberry fruits during shelf life. *LWT-Food Science and Technology, 85*, 440–444. Elsevier Ltd. doi: 10.1016/j.lwt.2016.12.056.

Maria, V., Clara, P., Amparo, C., Julian, C. D., & Lez-Marti΄Nez, C. G., (2008). Recent advances in edible coatings for fresh and minimally processed fruits. *Critical Reviews in Food Science and Nutrition, 48*, 496–511.

Mart, O., (2008). Edible coatings with antibrowning agents to maintain sensory quality and antioxidant properties of fresh-cut pears. *Postharvest Biology and Technology 50*, 87–94. doi: 10.1016/j.postharvbio.2008.03.005.

Martínez-Romero, D., Alburquerque, N., Valverde, J. M., Guillén, F., Castillo, S., Valero, D., & Serrano, M., (2006). Postharvest sweet cherry quality and safety maintenance by aloe vera treatment: A new edible coating. *Postharvest Biology and Technology, 39*(1), 93–100. Elsevier. doi: 10.1016/J.POSTHARVBIO.2005.09.006.

Martínez-Romero, D., Castillo, S., Guillén, F., Díaz-Mula, H. M., Zapata, P. J., Valero, D., & Serrano, M., (2013). Aloe vera gel coating maintains the quality and safety of ready-to-eat pomegranate arils. *Postharvest Biology and Technology*, *86*, 107–112. Elsevier B.V. doi: 10.1016/j.postharvbio.2013.06.022.

McHugh, T. H., & Senesi, E., (2000). Apple wraps: A novel method to improve the quality and extend the shelf life of fresh-cut apples. *J. Food Sci., 65*, 480–485.

Mei, Y., & Zhao, Y., (2003). Barrier and mechanical properties of milk protein-based edible films containing nutraceuticals. *J. Agric. Food Chem.*, 51, 1914–1918.

Moreíra, M. R., Tomadoni, B., Martín-Belloso, O., & Fortuny, R. S., (2015). Preservation of fresh-cut apple quality attributes by pulsed light in combination with gellan gum-based prebiotic edible coatings. *LWT-Food Science and Technology, 64*(2), 1130–1137. doi: 10.1016/j.lwt.2015.07.002.

Ortega-Toro, R., Collazo-Bigliardi, S., Roselló, J., Santamarina, P., & Chiralt, A., (2017). Antifungal starch-based edible films containing Aloe vera. *Food Hydrocolloids, 72*, 1–10. doi: https://doi.org/10.1016/j.foodhyd.2017.05.023.

Perez-Gago, M. B., Serra, M., Alonso, M., Mateos, M., & Del, R. M. A., (2005). Effect of whey protein- and hydroxypropyl methylcellulose-based edible composite coatings on color change of fresh-cut apples. *Postharvest Biology and Technology, 36*(1), 77–85. doi: https://doi.org/10.1016/j.postharvbio.2004.10.009.

Phan, T. D., Debeaufort, F. D., & Voilley, A., (2008). Moisture barrier wetting and mechanical properties of shellac/agar or shellac/cassava starch bilayer and biomembrane for food application. *Jour. of Membrane Sci., 325*, 277–283.

Qui, M., Jiang, H., Ren, G., Huang, J., & Wang, X., (2012). Effect of chitosan coatings on postharvest green asparagus quality. *Carbo-Hydrate Polymers, 92*(2), 2027–2032.

Raghav, P. R., Nidhi, A. N., & Saini, M., (2016). Edible coating of fruits and vegetables: A review *International Journal of Scientific Research and Modern Education (IJSRME), I*(I). ISSN: Online: 2455 – 5630. www.rdmodernresearch.com (accessed on 19 December 2020).

Raybaudi-Massilia, R. M., Mosqueda-Melgar, J., & Martín-Belloso, O., (2008). Edible alginate-based coating as carrier of antimicrobials to improve shelf life and safety of fresh-cut melon. *International Journal of Food Microbiology, 121*(3), 313–327. doi: https://doi.org/10.1016/j.ijfoodmicro.2007.11.010.

Robertson, G. L., (2009). *Food Packaging, Principle, and Practices* (2nd edn.).

Robertson, G. L., (2013). *Food Packaging, Principle and Practices* (3rd edn.) https://www.crcpress.com/Food-Packaging-Principles./Robertson/./9781439862414 (accessed on 19 December 2020).

Robles-Sánchez, R. M., Rojas-Graü, M. A., Odriozola-Serrano, I., González-Aguilar, G., & Martin-Belloso, O., (2013). Influence of alginate-based edible coating as carrier of antibrowning agents on bioactive compounds and antioxidant activity in fresh-cut Kent mangoes. *LWT-Food Science and Technology, 50*(1), 240–246. Elsevier Ltd. doi: 10.1016/j.lwt.2012.05.021.

Romanazzia, G., Erica, F., Silvia, B. B., & Dharini, S., (2017). Shelf-life extension of fresh fruit and vegetables by chitosan treatment. *Critical Reviews in Food Science and Nutrition, 57*(3), 579–601.

Salvia-Trujillo, L., Rojas-Graü, M. A., Soliva-Fortuny, R., & Martín-Belloso, O., (2015). Use of antimicrobial nanoemulsions as edible coatings: Impact on safety and quality attributes of fresh-cut Fuji apples. *Postharvest Biology and Technology, 105*, 8–16. Elsevier B.V. doi: 10.1016/j.postharvbio.2015.03.009.

Sanchís, E., González, S., Ghidelli, C., Sheth, C. C., Mateos, M., Palou, L., & Pérez-Gago, M. B., (2016). Browning inhibition and microbial control in fresh-cut persimmon (*Diospyros kaki* Thunb. cv. Rojo Brillante) by apple pectin-based edible coatings. *Postharvest Biology and Technology, 112*, 186–193. Elsevier B.V. doi: 10.1016/j.postharvbio.2015.09.024.

Sharma, S., & Rao, T. V. R., (2015). Xanthan gum-based edible coating enriched with cinnamic acid prevents browning and extends the shelf life of fresh-cut pears. *LWT-Food Science and Technology, 62*(1), 791–800. Elsevier Ltd. doi: 10.1016/j.lwt.2014.11.050.

Singh, S., Khemariya, P., Rai, A., Rai, A. C., Koley, T. K., & Singh, B., (2016). Carnauba wax-based edible coating enhances shelf life and retain the quality of eggplant (*Solanum melongena*) fruits. *LWT-Food Science and Technology, 74*, 420–426. Elsevier Ltd. doi: 10.1016/j.lwt.2016.08.004.

Supapvanich, S., Prathaan, P., & Tepsorn, R., (2012). Postharvest biology and technology browning inhibition in fresh-cut rose apple fruit cv. *Taaptim jaan* using konjac glucomannan coating incorporated with pineapple fruit extract. *Postharvest Biology and Technology, 73*, 46–49. Elsevier B.V. doi: 10.1016/j.postharvbio.2012.05.013.

Synowiec, A., Gniewosz, M., Kraśniewska, K., Przybył, J. L., Bączek, K., & Węglarz, Z., (2014). Antimicrobial and antioxidant properties of pullulan film containing sweet basil extract and an evaluation of coating effectiveness in the prolongation of the shelf life of apples stored in refrigeration conditions. *Innovative Food Science and Emerging Technologies, 23*, 171–181. doi: https://doi.org/10.1016/j.ifset.2014.03.006.

Tapia, M. S., Rodrı, F. J., Carmona, A. J., & Martin-belloso, O., (2007). *Alginate and Gellan-Based Edible Coatings as Carriers of Anti-browning Agents Applied on Fresh-Cut Fuji Apples, 21,* 118–127. Article in Press. doi: 10.1016/j.foodhyd.2006.03.001.

Tapia, M. S., Rojas-Graü, M. A., Carmona, A., Rodríguez, F. J., Soliva-Fortuny, R., & Martin-Belloso, O., (2008). Use of alginate- and gellan-based coatings for improving barrier, texture, and nutritional properties of fresh-cut papaya. *Food Hydrocolloids, 22*(8), 1493–1503. doi: https://doi.org/10.1016/j.foodhyd.2007.10.004.

Valverde, J. M., Valero, D., Martínez-Romero, D., Guillén, F., Castillo, S., & Serrano, M., (2005). Novel edible coating based on aloe vera gel to maintain table grape quality and safety. *Journal of Agricultural and Food Chemistry, 53*(20), 7807–7813. American Chemical Society. doi: 10.1021/jf050962v.

Vieira, J. M., Flores-López, M. L., De Rodríguez, D. J., Sousa, M. C., Vicente, A. A., & Martins, J. T., (2016). Effect of chitosan-Aloe vera coating on postharvest quality of blueberry (*Vaccinium corymbosum*) fruit. *Postharvest Biology and Technology, 116*, 88–97. Elsevier B.V. doi: 10.1016/j.postharvbio.2016.01.011.

Xing, Y., Li, X., Xu, Q., Jiang, Y., Yun, J., & Li, W., (2010). Effects of chitosan-based coating and modified atmosphere packaging (MAP) on browning and shelf life of fresh-cut lotus root (*Nelumbo nucifera* Gareth). *Innovative Food Science and Emerging Technologies, 11*(4), 684–689. Elsevier Ltd. doi: 10.1016/j.ifset.2010.07.006.

Zapata, P. J., Navarro, D., Guillén, F., Castillo, S., Martínez-Romero, D., Valero, D., & Serrano, M., (2013). Characterization of gels from different Aloe spp. as antifungal treatment: Potential crops for industrial applications. *Industrial Crops and Products, 42*, 223–230. Elsevier. doi: 10.1016/J.INDCROP.2012.06.002.

Zhang, L., Li, S., Dong, Y., Zhi, H., & Zong, W., (2016). Tea polyphenols incorporated into alginate-based edible coating for quality maintenance of Chinese winter jujube under ambient temperature. *LWT-Food Science and Technology, 70*, 155–161. Elsevier Ltd. doi: 10.1016/j.lwt.2016.02.046.

CHAPTER 6

ACTIVE PACKAGING OF FRUITS AND VEGETABLES: QUALITY PRESERVATION AND SHELF-LIFE ENHANCEMENT

GAURAV KR DESHWAL,[1] SWATI TIWARI,[2]
NARENDER RAJU PANJAGARI,[1] and SAHAR MASUD[3]

[1]*ICAR-National Dairy Research Institute, Karnal, Haryana, India, E-mail: ndri.gkd@gmail.com (G. K. Deshwal)*

[2]*Department of Food Science and Technology, Pondicherry University, Puducherry, Tamil Nadu, India*

[3]*Division of Livestock Production and Management, Sher-e-Kashmir University of Agricultural Sciences and Technology of Jammu, Jammu and Kashmir, India*

ABSTRACT

The demand for fruits and vegetables has shown a rapid upsurge owing to their several health benefits. Various factors like oxygen, storage temperature, microbial decay, respiration rate and presence of ethylene may affect the shelf-life and quality of fruits and vegetables. Increased export and import, demand for fresh, safe and nutritious fruits and vegetables have evolved innovative packaging technologies to meet these requirements. Active packaging involves the incorporation of certain constituents into the packaging material or package headspace to enhance the shelf-life and quality of the package content. This book chapter focuses on the active packaging systems for enhancing the shelf-life of fruits and vegetables. Among these, ethylene absorbers, antimicrobial releasing agents and oxygen scavengers have shown promising results. Along with, commercially available and patented active packaging systems for fruits and vegetables are presented in tabular form.

6.1 INTRODUCTION

Packaging has become the third-largest industry in the world, which is next to the food and petroleum industry and represents a major portion of every nation's economy. Packaging functions have been extended from simple preservation to convenience, marketing, material reduction, safety, tamper-evidence, and eco-friendly nature (Corrales et al., 2014). In this perspective, advanced packaging techniques such as modified atmosphere packaging (MAP), active packaging, and intelligent packaging are growing at a very high rate. The change in the gas composition of the atmosphere surrounding the product followed by sealing in high moisture and gas barrier packaging material is known as MAP. Vacuum packaging (VP) is also a form of MAP as the atmosphere inside the product is modified by removing most of the air (Kerry et al., 2006). Many products are subjected to MAP having reduced oxygen in the headspace to increase the shelf life. Products packed under MAP include cured meat, fruits, vegetables, fresh pasta, bakery products like biscuits and cookies, poultry, cheese, etc.

Consumer inclination towards safe and healthy food have led to the development of state-of-the-art and unique approaches in food processing and packaging. One such development is the smart packaging technology. Smart packaging should not be confused with intelligent packaging, as they both are different terms. However, in conferences and symposiums, speakers are using the term smart and intelligent packaging interchangeably. Intelligent packaging is the system which senses and indicates, while smart packaging is the combination of active and intelligent packaging properties (Yam et al., 2005). Active packaging is the incorporation of constituents into the packaging system for enhancement of package performance in terms of improved product characteristics or its shelf life. Considering the viewpoint of horticultural produce, factors affecting the quality of fruits and vegetables (F&V) like physiological processes (e.g., respiration rate of horticultural produce), chemical processes, physical processes (e.g., dehydration), infestation by insects and microbiological spoilage need to be addressed by active packaging technologies.

Active packaging has witnessed significant growth due to newly developed products and their packaging requirements along with consumer preferences over the past decade. The total market value of packaging in the United States was over $130 billion in 2008 and 55–65% of the total market, i.e., $64 billion was represented by active, intelligent, and MAP of foods and beverages (Lord, 2008). Active packaging market was worth

$6.4 billion segment in 2013 and is expected to touch the figures of $8.2 billion in 2018. Oxygen scavengers and moisture absorbers are the most commercialized forms of active packaging. Gas scavengers were the leading marketed product in 2012 in the USA (Realini and Marcos, 2014). However, strict laws for food safety being imposed by the European Union (EU) seem to be a hurdle in the commercialization of technologies based on smart packaging. In the coming future, smart packaging (active and intelligent) will be having very high potential as a technical tool in conveying the information to consumers along with shelf-life enhancement for producers (addressing issues of horticultural produce during transportation and storage) and processors of F&V.

6.2 HISTORY OF ACTIVE PACKAGING

Dr. Theodore Labuza from the University of Minnesota presented a groundbreaking review at EU conference in 1987 at Ireland and used the term 'active packaging' (Labuza and Breene, 1989). Active packaging in the form of advanced packaging which helps in improving the shelf life, quality, and safety of the product by either reacting with the product, headspace of the package or external environment. Some of the important active packaging systems are oxygen scavengers, carbon dioxide absorbers/emitters, moisture absorbers, self-heating, and self-cooling containers, antimicrobial packaging, ethanol emitters, flavor absorbers/releasers and microwave-assisted containers (Ozdemir and Floros, 2004). The first reported antimicrobial film was developed in 1954 for natural cheese by spraying sorbic acid over a wrapper. The first patent for oxygen absorber using enzyme glucose oxidase was created in 1956, while the use of sodium carbonate was mentioned in a German patent during 1968. Researchers in Australia and the USA published results regarding the usage of potassium permanganate as ethylene scavenger in bananas during 1970. However, all these applications were not instantly commercialized, but they laid foundation for evolvement of active packaging systems and first commercialized application of active packaging was in year 1976 in Japan, where iron powder sachets were used for oxygen scavenging (Robertson, 2013). In the near future, the active packaging market will further be stimulated by new forms of active packaging like microwave susceptor packaging, changing gas permeability, and self-venting packaging. However, factors like cost; performance under adverse conditions, legal issues will still be limitations (Realini and Marcos, 2014).

6.3 NEED OF ACTIVE PACKAGING FOR FRUITS AND VEGETABLES (F&V) IN DEVELOPING COUNTRIES

In the recent past, the need for active packaging for horticultural produce was motivated due to huge losses caused by post-harvest disorders. Other factors driving development of active packaging includes marketing requirement of better cold-chain due to evolvement of supermarkets and e-commerce, social changes like retail packs, home delivery, bulk shopping, fresh-cut F&V and technological advances in material science, nanotechnology, and informatics. Various factors which focus on the requirement of active packaging for F&V are:

- Absence of cold chains during post-harvest handling generates the need of packaging interventions which may preserve the product for a longer duration.
- Lack of well-designed quality control systems, especially in developing countries.
- Logistic supply changes with longer transportation distances and exports require well-established technologies for food safety and quality control.
- Emerging spoilage causing and pathogenic microflora in F&V prerequisites high-end control mechanism without having adverse effects on horticultural produce quality.
- Strict regulations regarding additive incorporation in F&V favors alternative tactics such as active packaging.
- Increased demand of processed F&V with low thermal treatment and enhanced shelf life (e.g., minimally processed F&V).

6.4 ACTIVE PACKAGING SYSTEMS FOR FRUITS AND VEGETABLES (F&V)

The focus of this section is to review the concept of active packaging systems and their application in F&V with emphasis on recent developments and commercialized technologies. Various types of active packaging technologies used in F&V along with their commercially available products are detailed in Table 6.1.

6.4.1 ETHYLENE ABSORBERS

Ethylene is sometimes referred to as the ripening or death hormone because it mainly accelerates the respiration rate and results in subsequent senescence (aging) of F&V. Various positive effects of ethylene include flowering in pineapples, color development in citrus fruits, bananas, and tomatoes and stimulation of root production in baby carrots (Abeles et al., 1992). Ethylene can also cause a range of postharvest physiological disorders such as russet spotting on lettuce, scald on apples and bitter flavor in cucumbers. Therefore, it is desirable to remove ethylene or to suppress its negative effects in most horticultural produce. To this aim incorporating ethylene scavengers into fresh produce packaging and storage areas has been the focus of much research.

TABLE 6.1 Summary of Commercially Available Active Packaging Systems Used in Fruits and Vegetables

Active Packaging Systems Used in Fruits and Vegetables	Active Components/Interventions Used	Commercial Brand Names
Ethylene absorbers	Potassium permanganate, Activated carbon, Zeolite	Neupalon™ Peakfresh™ Evert-Fresh™
Oxygen scavengers	Iron salts, Ascorbate salts, Enzymes (glucose oxidase and catalase)	Ageless® Freshpax® AegisOX®
Carbon dioxide scavengers	Calcium hydroxide, Magnesium oxide, Active charcoal, Zeolite	Emco® Lipman® Ageless®
Moisture absorbers	Silica gel, Calcium oxide, Calcium chloride, Modified starch, Natural clays (e.g., montmorillonite), Humectants (Propylene glycol).	Peaksorb® Pichit™ Dri-loc®
Antifogging agents	Biaxially oriented vinylon, Compression rolled oriented HDPE, Hydrophilic polymers, Micro-perforations	Visgard® Clarifoil®
Antimicrobial agents	Silver ions, Chlorine dioxide, Sulfur dioxide, Essential oils (e.g., eugenol, thymol, carvacrol, menthol, eucalyptol), Spices, Herbs, Organic acids (sorbate, propionate, and benzoate), Bacteriocins (nisin), Enzymes (lysozymes), and Antioxidants	GrapeGuard® EthylBloc™ SmartFresh™ AgroFresh®
Odor control	Films containing ferrous salt and ascorbate	Anico™ bags MINIPAX® ScentSational®
Changing gas permeability	Micro-perforations, Hydrosorbent, Side chain crystallizable polymers	BreatheWay® Intellipac®

The ethylene absorbers are the most researched and utilized category of active packaging in F&V. The most commonly used ethylene absorbers for F&V include potassium permanganate ($KMnO_4$), activated carbon, and zeolites. Most of the commercialized ethylene scavengers based on potassium permanganate reaction involves ethylene (C_2H_4) oxidation to acetaldehyde, acetic acid and finally to carbon dioxide (CO_2) and water (H_2O). However, owing to the toxic nature of $KMnO_4$, it is added to inert substances with very high surface area such as perlite or silica gel and stored in a sachet. Granular activated carbon (GAC) impregnated with palladium (Pd) as catalyst had been reported to possess better ethylene removal capacity as compared to activated carbon. However, the higher cost of palladium limits its applicability. GAC with 1% Pd system had been tested with modified atmosphere packaged tomatoes for its ethylene scavenging efficiency (Bailen et al., 2013). The results showed delay in tomato ripening, reduced weight loss, and enhanced shelf life without any decay and off-flavor. In another study, broccoli stored in ethylene scavenging linear density polyethylene (LDPE) bags at 10°C showed two days increase in shelf life (Jacobsson et al., 2004).

6.4.2 PRESERVATIVE OR ANTIMICROBIAL AGENT RELEASERS

The surface microbial growth and contamination during harvesting, processing, and transportation is the major cause of F&V deterioration. Freshly harvested F&V are contaminated with mixed microflora consisting of *E. coli*, *Pseudomonas*, and *Erwinia* whilst, fresh-cut F&V when stored aerobically under refrigerated condition majorly contain yeasts, molds, and *Pseudomonas* (May and Fickak, 2003). The washing of F&V using hydrogen peroxide, peroxyacetic acid, ozone, and chlorinated water can reduce microbial count, but total elimination will not be achieved (Akbas and Olmez, 2007). The remaining microflora in F&V generates the requirement of antimicrobial packaging. Preservative releasers and antimicrobial-based packaging can be defined as the interaction of active ingredients with food product, which enables inhibition of spoilage and pathogenic microorganisms (Suppakul et al., 2003). Mainly two terms in the above definition "active ingredient" and "interaction" needs to be more clearly defined with reference to F&V. Firstly, active ingredients used for F&V mainly consists of silver ions, chlorine dioxide, sulfur dioxide, essential oils (e.g., eugenol, thymol, carvacrol, menthol, eucalyptol), spices, herbs, organic acids (sorbate, propionate, and benzoate), bacteriocins (nisin), enzymes (lysozymes) and antioxidants. Secondly, the interaction between horticultural produce an antimicrobial agent can be brought by following ways: release of antimicrobial in package headspace; antimicrobial incorporated in packaging material and released during storage; immobilization of antimicrobial on package surface; and package material with antimicrobial properties can be utilized for packaging (Vermeiren et al., 1999). The active ingredients having high heat resistance can be incorporated into films using extrusion and injection molding, while heat labile constituents assimilated using solvent compounding, coating, and casting.

Chlorine dioxide (ClO_2) is bactericidal gas with a broad-spectrum and used worldwide for treatment of drinking water and cleaning of F&V. Chlorine dioxide containing films with its controlled release, for packaging of F&V are available commercially worldwide (Scully and Horsham, 2007). The fumigation with sulfur dioxide (SO_2) gas in cold stores is mainly used to prevent the growth of *Botrytis cinerea* fungus, thus controlling gray mold on table grapes (Mustonen, 1992). However, during later storage also this defect may occur, thus reflecting the need for controlled release of SO_2. A Chile firm (Quimica OSKU S.A.) developed OSKU GrapeGuard® pads for release of SO_2 using reaction between water and sodium metabisulfite resulting in 8 to 10 weeks shelf-life of table grapes at 0°C. EthylBloc™ and SmartFresh™ developed by AgroFresh® for release of 1-MCP (methylcyclopropene) which had been reported to enhance shelf-life of various F&V even at very low doses (Scully and Horsham, 2007). Chitosan is mainly used as a carrier of antimicrobial additives. Chitosan as an antimicrobial agents had been studied for fresh-cut papaya, tomato, strawberries, radish, kiwifruit, etc., which resulted in reduced microbial growth and enhanced shelf life. Moreover, the decay in fresh melons caused by spoilage fungi was considerably reduced by natamycin in a bilayer coating of chitosan (Cong et al., 2007). In addition, the natural compounds like hexenal and hexyl acetate had also been used for enhancing the shelf life of minimally processed fruits (Lanciotti et al., 2004).

Essential oils (volatile or ethereal oil) are aromatic natural compounds derived from plant organs like bud, seed, leave, twig, bark, root, and flower. Owing to their natural origin, fragrance, flavor, and antimicrobial properties there use is gaining importance (Fisher and Phillips, 2008). Alginate coatings containing essential oils like lemongrass, oregano, and vanillin on cut apples decreased the growth of psychrophilic aerobes, yeasts, and molds by 2 log cfu/g (Rojas-Grau et al., 2009). Table grape and sweet cherry shelf-life was enhanced significantly with reduced softening in both fruits because of essential oil (eugenol, thymol, menthol) addition to MAP packages in comparison to control MAP (Serrano et al., 2008). Essential oils (e.g., eugenol, and thymol) had also been reported to reduce the loss of phenolics and anthocyanins from skin and ascorbic acid from flesh of grapes (Serrano et al., 2005). Visible fungal growth of *Candida albicans*, *Aspergillus flavus*, and *Eurotium repens* in strawberry was completely inhibited for 7 days at 4°C using clove and other essential oils in the form of paper coatings (Dini, 2016).

6.4.3 OXYGEN SCAVENGERS

The presence of oxygen can cause various undesirable changes in F&V namely: ascorbic acid or vitamin C loss, growth of aerobic spoilage microorganisms, acceleration of fresh F&V respiration, enzymatic, and non-enzymatic phenolic browning of fresh fruit and discoloration of processed fruit and vegetables pigment. Removal of oxygen has been the prime focus of food industries using MAP, VP, or high barrier packaging materials (Polyakov and Miltz, 2010). Despite, complete removal of oxygen from fruit and vegetable packages is not well achieved. Therefore, oxygen scavengers can be highly beneficial for removal of oxygen from the food package's headspace after MAP or oxygen permeating through packaging material or oxygen remaining between F&V slices. Oxygen scavengers, also referred to as oxygen absorbers are based on iron salts, ascorbate salts, and enzymes like glucose oxidase and catalase, and available in various forms like sachets and pads. The oxygen scavengers based on iron (Fe) salts absorbs oxygen (O_2) in the presence of moisture (H_2O) and its activity is based on the following reaction:

$$Fe + \frac{3}{4}O_2 + \frac{3}{2}H_2O \rightarrow Fe(OH)_3$$

The use of iron-based oxygen scavengers provide the advantage of reducing headspace oxygen to less than 0.01%, which is very low in comparison to residual oxygen level of 0.3 to 3% in vacuum or MAP. The oxygen removal rate of iron powder is well postulated, with 1 g of iron powder removing 300 mL of oxygen which is equivalent to 0.0136 mol of oxygen (standard temperature and pressure) (Robertson, 2013). Recently, pyrogallol coated LDPE films were found effective in reducing oxygen concentration from 20.9% to 9.42% at 23°C with water working as an activator for its oxygen scavenging activity (Gaikwad et al., 2017).

Charles and co-workers demonstrated rapid depletion of the oxygen level in the headspace of tomatoes and mushrooms using iron-based oxygen-scavenging sachets, thereby resulting in a 50% reduction in the time required to achieve equilibrium oxygen and carbon dioxide concentration. The harmful transient peak in carbon dioxide concentration observed in the absence of the sachet was almost completely eliminated through incorporation of a sachet in the package (Charles et al., 2003). However, careful synchronization of oxygen-scavenging rate and capacity with respiration characteristics of the produce and permeation properties of the packaging material is required to avoid the emergence of anaerobic condition (Charles et al., 2006). The use of oxygen scavengers (ATCO®-100 having 100 mL scavenging capacity and ATCO®-210 having 210 mL scavenging capacity) increased the shelf life of fresh strawberries to 4 weeks and also resulted in lower mold growth, lower loss of total soluble solid, better color and higher sensory scores (Aday and Caner, 2013). Commercially available oxygen scavenger (Ageless Z*, Cryovac, Australia) disinfested *Tribolium castaneum* in non-MAP consumer packages of dried raisins and also extended their shelf life without having any negative effect of O_2 scavenger (Tarr and Clingeleffer, 2005).

6.4.4 CARBON DIOXIDE SCAVENGERS AND EMITTERS

Carbon dioxide's excessive production during respiration of fresh F&V causes physiological stress in horticultural produce, ballooning, and even bursting of packages. Therefore, controlled removal of CO_2 can help in maintaining the quality of the product. Calcium hydroxide and magnesium oxide are the most commonly used chemical CO_2 absorbents. Physical absorbents like active charcoal and zeolite work well only in low humidity conditions because of their greater affinity to moisture than to CO_2.

Excess carbon dioxide produced during respiration of fruits or vegetables is removed by these physical and chemical absorbents, thus overcoming the adverse effects such as reduced pH, color, and flavor changes (Imran et al., 2010). Kimchi is a general term for fermented vegetables such as oriental cabbage, radish, green onion, and leaf mustard mixed with salt and spices, and it is a CO_2-producing food product. Pasteurization or any kind of thermal treatment of kimchi leads to loss of its sensory qualities furthermore, parallel production of CO_2 during fermentation process created the need for CO_2 scavengers in packed kimchi (Coles et al., 2003). Carbon dioxide scavengers had also been tested with strawberry resulting in reduced mold decay, retarded senescence, and better sensory properties (Almenar et al., 2007). Similarly, in pear and eggplant carbon dioxide scavengers prevented browning and reduced chilling injury incidences (Lee, 2016).

On the other side of the coin, a higher level of carbon dioxide in the headspace of packaged F&V can retard the growth of aerobic microorganisms, lower respiration and delay the aging process (Chakraverty, 2001). Carbon dioxide emitting materials are based on the reaction of sodium bicarbonate and water. However, the use of carbon dioxide emitting material can cause anaerobic metabolism in packed F&V thus, during designing of the package, film permeability and respiration rate should be considered as prime factors. Some of the commercially available CO_2 emitter are Verifrais™ (SARL Codimer, France), Freshpax®, Freshlock® (Multisorb Technologies, USA) and CO_2 Fresh Pads (CO_2 Technologies, USA). Carbon dioxide emitters and absorbers could be used as per the requirement, with the aim of having a balanced atmosphere in F&V packages. A brief summary of various types of active packaging systems used in different vegetables are presented in Table 6.2.

TABLE 6.2 Active Packaging Systems for Vegetables

Vegetables	Active Packaging Type	Concept/Active Ingredients	References
Broccoli	Ethylene absorbers	Linear density polyethylene (LDPE) films with Tazetut® ethylene absorber	Esturk et al. (2014)
Kimchi	CO_2 absorbers	Zeolite and active carbon sachets	Lee et al. (2001)
Carrot	Antimicrobial	Chitosan and sodium caseinate coatings	Moreira et al. (2011)
Green bell pepper	Moisture absorbers	Sachets of silica gel crystals	Singh et al. (2014)
Tomatoes	Oxygen absorbers	Iron-based scavenger	Charles et al. (2003)
Okra	Antimicrobial	Polyethylene films with chitosan-zinc oxide nanocomposites	Al-Naamani et al. (2018)
Kimchi	Changing gas permeability	Pinhole in package with gas flushing	Lee and Paik (1997)
Mushrooms	Moisture control	Humidity regulating trays with salt	Rux et al. (2015)

6.4.5 CHANGING GAS PERMEABILITY

Gas-permeable packaging allows gas exchange between the outside environment and inside of the package to ensure an equilibrium that best suits the packaged food product. Formerly, micro-perforations, and incorporation of inorganic particles into polyethylene have been used for the development of customized F&V packages with gas transmission ability. The major drawback of such packages was their inability to adjust gas transmission rate according to variation in respiration rate of horticultural produce. Subsequently, BreatheWay® membrane technology (Apio Inc., California), based on side-chain crystallizable (SCC) polymers provides the solution for gas permeability control according to change in temperature. SCC polymers become molten liquid at a particular switch temperature, thus enhancing the gas permeability. The change in polymer properties like chain length and side chains can be used for attaining required oxygen and carbon dioxide permeabilities in F&V packages (Robertson, 2013). The commercial uptake of SCC based technology is the adoption of Landec Corporation Intellipac® for distribution of bananas to superstores in the USA by Chiquita Brands International™ (Clarke, 2002). Further, coated hydrophobic polymers containing a high potency hydrosorbent allows the tracking and control of water vapor, oxygen, and carbon dioxide transfer rate (Mathlouthi et al., 1994). This system can work best for F&V that produce gases and moisture during processes like respiration, transpiration, and condensation.

6.4.6 HUMIDITY AND CONDENSATION CONTROLS

Package permeability and transpiration are the major cause of moisture presence in packages of F&V. The transpiration rate depends on temperature and also varies during day and night. Excess moisture produced by transpiration in F&V causes their spoilage of by accelerating various undesirable reactions, fungal, and bacterial growth, shriveling, and loss of sensory qualities (Chakraverty, 2001). Respiration of horticultural produce in high barrier packaging material leads to elevated relative humidity (RH) in the package, thus causing condensation of moisture and affecting both package appearance and product quality. Moisture absorbers, humidity regulators, and desiccants are used to soak up the moisture, thereby extending the shelf life by inhibiting microbial growth and texture degradation (Rico et al., 2007).

One of the earliest available plastic-based active moisture-absorbing packaging material was patented by Patterson and Joyce. It consisted of multilayered packaging material with innermost hydrophobic but water-permeable layer, middle layer of moisture absorbent material like cellulosic paper or polyvinyl alcohol (PVOH) and an outer layer of water-impermeable material (polyethylene). This form of liner bags were used in fiberboard cartons for F&V storage as it itself acts as a water absorber without any desiccants (Patterson and Joyce, 1994). Moisture absorbers are commercially available in the form of sachets, blankets, pads, and sheets. Moisture absorbing sachets are mainly used for dried foods while foods with high water activity like fruits, vegetables, fish, poultry, and meats are provided with drip absorbent pads, sheets, and blankets. Chemicals which are mainly used for moisture absorption includes silica gel, calcium oxide, calcium chloride, modified starch, and natural clays (e.g., montmorillonite). Fruits are often wrapped with moisture-absorbing papers. Microporous sachets of sodium chloride as desiccant have been used for tomato distribution in the United States of America (USA) (Day, 2008).

Moisture blocking in the vapor phase by reducing RH inside the pack had been a very good approach for high water activity food commodities like fruits, vegetables, and cheese, meat, and poultry products. It mainly involves one or more humectants (agents that reduce a_w) placed between two layers of water-permeable plastic film. A commercially available material based on similar concept was Pichit™ (Showa Denko Co. Ltd., Japan), which consisted of humectant propylene glycol sandwiched between PVOH layers. Pichit films are commercially available in the form of rolls for wrapping fruits, vegetables, fresh meats, fish, and poultry. This film had been reported to extend shelf life by 3–4 days as it reduces surface moisture by osmotic dehydration (Labuza and Breene, 1989). Polymers films and boxes with a definite number of micro-perforations for moisture control in F&V were introduced latterly. However, due to higher moisture release and its uneven distribution, this system did not prevented quality deterioration. Contrarily, a hollow fiber membrane contractor was developed by Dijkink et al. (2004) for control of RH at a level of 90% during bell pepper storage and reduction in fungal development without any shriveling even after 3 weeks.

Moisture presence in packed fruit and vegetable packages cause fog formation, which further leads to shabby condition and loss of product's appearance. Fogging is mostly seen in fresh fruits, vegetables, and meats. The difference between surface tension of packaging film (especially polyolefins like polyethylene or polypropylene (PP)) and water leads to condensation of moisture (fog) on the package's surface when internal air gets cooled to its dew point. Due to hydrophobic, non-polar nature and lower surface energy of polyolefins as compared to water, moisture condenses on the film's surface as droplets with high contact angles. This problem can be overcome by incorporating antifogging agents in films which includes humidity absorbers, hydrophilic liners, and micro-perforations in film. Antifogging films reduce internal vapor pressure, thus preventing moisture condensation in respiring F&V (Ozdemir and Floros, 2004).

6.4.7 ODOR CONTROLS

Off-flavor development is an important undesirable biochemical process occurring during MAP of F&V. Off flavor formation may serve as an indicator of spoilage or shelf-life determination of fresh-cut fruit and vegetable. An increase in biosynthesis of ethanol and acetaldehyde during MAP and fresh storage of F&V had been directly correlated with off-flavor development. Films containing ferrous salt and ascorbate were used to develop odor-absorbing bags (Anico™, Anico Co. Ltd., Japan) for removal of undesirable amines and aldehydes such as hexanal by their oxidation.

The use of flavor/odor releasers is controversial because of their ability to mask undesirable spoilage reactions, thus misguiding the consumers. Conversely, some products like onion and potatoes naturally possess a pungent odor. This creates the demand for odor absorbers as off-odor may cross-contaminate other commodities during their storage or distribution. Durian, which is considered as the king of fruits and smelliest fruit in Southeast Asia, had been banned from all international airlines owing to its typical off-flavor. The major requirement of active packaging for such types of fruits requires such packaging material which may prevent the passage of undesirable gases without disturbing the balanced surrounding condition of the package. Morris developed a packaging system with odor impermeable plastic (e.g., polyethylene or polyethylene terephthalate (PET)) for durian, whilst entailing a port for passage of respiratory gases of the fruit. This port was fixed with a sachet containing charcoal and 10% nickel for odor absorption (Rao and Srivastava, 2018). Different active packaging systems of fruits along with their reactive component are summarized in Table 6.3.

6.4.8 TEMPERATURE CONTROL PACKAGING

It is clearly known that temperature is the most important factor for microbial, respiration rate, and chemical reactions and has great influence on plant metabolic activity. The deterioration process of F&V can be retarded by precise control of temperature and as a general rule; each increase of 10°C increases the respiration rate of plants nearly twice (Atkin and Tjoelker, 2003). In addition, the increase of temperature leads to an exponential increase in microbiological growth and spoilage rate, until the temperature reaches a level at which the microorganisms may be thermally disrupted or destroyed. One suitable approach to prevent temperature changes in the produce is application of temperature control packaging. Currently, temperature controlled packaging in individual fruit and vegetable packages is at the budding stage, and more research is required in this element of active packaging by technology developers, which would lead to cost-effective temperature control of a single package.

TABLE 6.3 Active Packaging Systems for Fruits

Fruits	Active Packaging Types	Concept/Active Ingredients	References
'Fuji' apples	Antimicrobial	Polyvinyl chloride films coated with nano-zinc oxide powder	Li et al. (2011)
Cantaloupe melon	Antimicrobial	Vanillin and cinnamic acid vapors through a filter paper	Silveira et al. (2015)
Banana	Ethylene absorbers	Potassium permanganate	Chauhan et al. (2006)
	CO_2 absorbers	Soda-lime	
	Moisture absorbers	Silica gel	
Pineapple	Antimicrobial	Alginate based coating with essential oil lemongrass	Azarakhsh et al. (2014)
Oranges	Antifungal coatings	Chitosan and essential oils (bergamot, thyme, and tea tree oil)	Chafer et al. (2012)
Blueberries	Antimicrobial	Chlorine dioxide pads	Sun et al. (2014)
Wild Pomegranate	Moisture absorbers	Salt and sugar sachets	Sharma and Thakur (2016)
Kiwifruit	Antimicrobial	Polyethylene bags with zeolite	Sezer et al. (2017)

6.4.9 ACTIVE NANOCOMPOSITES FOR FRUITS AND VEGETABLES (F&V)

Materials with at least one dimension less than 100 nm are considered as nanomaterial. There are mainly two approaches for the production of nanomaterials: materials broken down to smaller particles physically or chemically is referred as "top-down approach" while "bottom-up" or "self-assembly" is an arrangement of molecules one at a time to build macro-sized structure (Warad and Dutta, 2005). Owing to high surface area to volume ratio, they are used in various food-packaging applications at very low concentration for improved material properties and better dispersion of active component. Active material in sachets or pads is gradually replaced by active nanocomposites for controlled release of active substance for product preservation. Active nanocomposites films incorporating oxygen scavengers, antioxidants, and antimicrobials had been successfully tested for various food commodities like meat, cheese, bread, fish, poultry, fruits, and vegetables (Moreira et al., 2011). However, the applications and techno-commerciality of nanostructured material is very limited in the packaging of horticultural products. Silver nanoparticles in polymer nanocomposites packaging material had been reported to destruct ethylene gas in fruit packages, thus slowing their ripening rate and extending shelf life. Similarly, the potential of titanium dioxide (TiO_2) coated nano-packaging for reducing accumulation of ethylene, acetaldehyde, and ethanol in horticultural produce packages had been indicated but intensity of radiation, frequency of light, shadow effect and adverse impact on food components need to be considered (Maneerat and Hayata, 2006a, b).

6.4.10 COMBINATION OF ACTIVE PACKAGING SYSTEMS FOR FRUITS AND VEGETABLES (F&V)

The combination of two or more active packaging systems can be employed to produce a kind hurdle effect against deteriorating components within an individual packaging unit. A patent of two-layered packaging material for moisture-sensitive products like cheese and bakery products was created (Marbler and Parmentier, 1999). The packaging material consisted of first functional layer (e.g., coated paper) for storing and releasing moisture, and second functional layer (e.g., plastic laminate) for controlling gas permeability as a function of moisture content. Similarly, for F&V, combined active packaging systems could be developed in the near future. Aday et al. (2011) studied the effect of CO_2 (EMCO-A® with 24% sodium carbonate, EMCO-B® having 20% sodium carbonate) and O_2 absorbers (ATCO®) on the quality of fresh strawberries and reported extension of shelf-life, slower respiration rate and delayed ripening. Combination of oxygen and ethylene scavengers along with perforations in thermocol trays containing peaches and wrapped with LDPE showed the optimistic effectiveness of active packaging systems on shelf life, decay, weight loss, and overall acceptability of peaches (Mir and Bandral, 2018).

Plastic materials having combined oxygen-scavenging/carbon dioxide releasing capabilities have also been explored at Food Science Australia. In this regard, ethylcellulose was used as the base substrate that contained the active agents. The oxygen-scavenging functionality of the polymer film was achieved using the visible-light-driven photo-oxidation of a sensitizer molecule, tetraphenylporphine, to produce short-lived and highly reactive singlet-oxygen molecules. The singlet-oxygen molecules produced in this reaction then proceed to react with furoic acid also incorporated in the film containing the sensitizer, leading to the formation of an endoperoxide derivative of furoic acid that releases carbon dioxide upon rearrangement. On exposure to light, oxygen in the package headspace can be rapidly replaced with carbon dioxide without affecting the total headspace volume. Similarly, packages with such interesting features can be further explored for F&V without deteriorating its quality by light exposure (Scully and Horsham, 2007). Other unifications that could be used for F&V include moisture absorbing pads with antimicrobial releasers or a combination of ethanol emitters with ethylene absorbing applications. The partnership of active and intelligent packaging can be used to complement each other's actions.

6.5 PERCEPTION OF CONSUMER NEEDS BY DEVELOPER OF FRUITS AND VEGETABLES' (F&V) ACTIVE PACKAGING

As per active packaging developing industries' perspective, the development of active packaging system for F&V requires the identification and justification of consumer's need. Nevertheless, the major concern here is the identification of the real consumer who will judge the efficiency of active packaging intervention. The identification of real customer and their need helps in developing a superior quality product. Critically, major hypothesis in success of any new product or intervention is satisfaction of end-users which makes everyone happy. For example, a firm producing active packaging system for moisture absorption in fruit's package and selling their product to the packaging firm need to satisfy not only packaging firm but also supermarket, grocery seller, buyer, and consumer. Various consumer trends, facts, and surveys for identifying the needs of individual and firm could be helpful in designing an active packaging system for horticultural produce. The needs of consumer are to be surveyed and arranged according to their importance, which allows manufacturers to plan their research accordingly.

6.6 REGULATORY ISSUES OF ACTIVE PACKAGING

Some food safety and regulatory issues in relation to active packaging of F&V like approval for food contact active packaging systems, environmental regulations about active-packaging materials, evasion of any consumer confusion about package labeling by active packaging and adverse effect of active packaging on the microbial ecosystem should be considered (Rooney, 1995). Active packaging material in contact with food material may migrate into the food product and can have adverse health impacts, which creates demand for food contact approval and possible toxicological studies regarding the migrants. Environmental regulations involve the studies regarding reusability, recyclability, and energy recovery from the active components. Various warning and instructions should be provided on the consumer's pack to avoid ingestion of the active components. Food manufacturers should consider the effects of active packaging on the microbial and safety aspects of foods. For example, using oxygen scavengers in food packaging may remove all the O_2 from the packs of chilled perishable food products with high water activity and stimulate the growth of anaerobic pathogenic bacteria such as *Clostridium botulinum* (Betts, 1996). It is important in antimicrobial films that the spectrum of inhibited microorganisms must be determined. A number of patented and ultramodern active packaging technologies for their use in maintaining the quality, safety, and shelf life of horticultural commodities are discussed briefly in Table 6.4.

TABLE 6.4 Patented Literature of Fruits and Vegetables' Active Packaging Technologies

Title of Patent	Active Packaging Role/ Concepts	References
Packaging containing fragrance	Peppermint flavor packages for Royal Gala apples	Popplewell and Henson (2004)
Germicide capable of preventing apples from putrefaction and its use	Preventing apple canker	Chinese Patent No. CN1040126A (1988)
Method for obtaining an absorber ethylene and paper product obtained	Ethylene absorbing paper	Antonio (2018)
Multilayer oriented antimicrobial and antifogging films	Coextruded and oriented films with at least three layers	Aral et al. (2005)
Container comprising uniaxial polyolefin/filler films for controlled atmosphere packaging	Controlled flux of CO_2 and O_2 for retarding maturation of fruits and vegetables	Antoon (1990)
Packaging system for the control of relative humidity of fresh fruits, vegetables, and flowers with simultaneous regulation of CO_2 and O_2	Three-layered water-permeable sealed containers	Lim and Lee (2010)
Controlled-bactericide-release fruit and vegetable preservative packaging composite film	Film containing vinyl acetate, bactericide, and sulfur dioxide in three different layers	Chinese Patent No. CN101601420A (2009)

6.7 FUTURE TRENDS FOR ACTIVE PACKAGING OF FRUITS AND VEGETABLES (F&V)

Consumer preferences for naturally preserved and minimally processed food product along with more investment on food product's quality and safety had given a break to active packaging which is expected to rise exponentially in upcoming future. However, more studies regarding the impact of sachets, pads, labels, and other active ingredient's incorporation into food packages along with their mode of action are required. In addition, the studies should focus more attention on the function of preservatives where microbial growth and spoilage mainly occurs on the surface of food and develop new antimicrobial packaging materials that are effective against several spoilage and pathogenic microbes. It is understandable that these novel materials should have good appearance, good mechanical properties, proper permeability properties, be reasonable in price, and be suitable both for packaging machines already used in the food industry and for normal sealing procedures (Wilson, 2007). Research into the application of nanocomposites in active packaging should be focused on antioxidant-releasing films, light-absorbing/regulating systems, antifogging, and anti-sticking films, gas permeable/breathable films, bioactive agents for controlled release, and insect-repellent packaging. In context to respiring horticultural produce, synchronization of active component's release with product requirement is a prerequisite. An increasing demand for packaged fresh-cut horticultural produce generates future scope and challenges for active packaging applications.

KEYWORDS

- **carbon dioxide**
- **chlorine dioxide**
- **ethylene**
- **European Union**
- **granular activated carbon**
- **linear density polyethylene**
- **methyl cyclopropane**

REFERENCES

Abeles, F. B., Morgan, P. W., & Saltveit, Jr. M. E., (2012). *Ethylene in Plant Biology*. Academic Press.

Aday, M. S., & Caner, C., (2013). The shelf-life extension of fresh strawberries using an oxygen absorber in the biobased package. *LWT-Food Science and Technology*, *52*(2), 102–109.

Aday, M. S., Caner, C., & Rahvali, F., (2011). Effect of oxygen and carbon dioxide absorbers on strawberry quality. *Postharvest Biology and Technology*, *62*(2), 179–187.

Akbas, M. Y., & Ölmez, H., (2007). Effectiveness of organic acid, ozonated water, and chlorine dippings on microbial reduction and storage quality of fresh-cut iceberg lettuce. *Journal of the Science of Food and Agriculture*, *87*(14), 2609–2616.

Almenar, E., Del, V. V., Catala, R., & Gavara, R., (2007). Active package for wild strawberry fruit (*Fragaria vesca* L.). *Journal of Agricultural and Food Chemistry*, *55*(6), 2240–2245.

Al-Naamani, L., Dutta, J., & Dobretsov, S., (2018). Nanocomposite zinc oxide-chitosan coatings on polyethylene films for extending storage life of okra (*Abelmoschus esculentus*). *Nanomaterials*, *8*(7), 479.

Antonio, F. D. L. C., (2018). Spanish patent Application No. L ES 2685994A1. Method for obtaining an absorber ethylene and paper product.

Antoon, Jr. M. K., (1990). U.S. Patent No.: 4,923,703. Container comprising uniaxial polyolefin/ filler films for controlled atmosphere packaging. U.S. Patent and Trademark Office, Washington, DC.

Aral, O., Rzaev, Z., & Buyukakinci, C., (2005). *U.S. Patent No. 6,838,186*. Multilayered oriented antimicrobial and antifogging films. U.S. Patent and Trademark Office, Washington, DC.

Atkin, O. K., & Tjoelker, M. G., (2003). Thermal acclimation and the dynamic response of plant respiration to temperature. *Trends in Plant Science*, *8*(7), 343–351.

Azarakhsh, N., Osman, A., Ghazali, H. M., Tan, C. P., & Adzahan, N. M., (2014). Lemongrass essential oil incorporated into alginate-based edible coating for shelf-life extension and quality retention of fresh-cut pineapple. *Postharvest Biology and Technology*, *88*, 1–7.

Bailen, G., Guillen, F., Castillo, S., Zapata, P. J., Serrano, M., Valero, D., & Martinez-Romero, D., (2013). Use of a palladium catalyst to improve the capacity of activated carbon to absorb ethylene, and its effect on tomato ripening. *Spanish Journal of Agricultural Research*, *5*(4), 579–586.

Betts, G. D., (1996). *Code of Practice for the Manufacture of Vacuum and Modified Atmosphere Packaged Chilled Foods with Particular Regards to the Risks of Botulism*. Guideline No. 11, Campden and Chorleywood Food Research Association, Chipping Campden, Glos, UK.

Chafer, M., Sanchez-Gonzalez, L., Gonzalez-Martínez, C., & Chiralt, A., (2012). Fungal decay and shelf life of oranges coated with chitosan and bergamot, thyme, and tea tree essential oils. *Journal of Food Science*, *77*(8), E182–E187.

Chakraverty, A., (2001). *Postharvest Technology*. Enfield, NH, Scientific Publishers.

Charles, F., Sanchez, J., & Gontard, N., (2003). Active modified atmosphere packaging of fresh fruits and vegetables: Modeling with tomatoes and oxygen absorber. *Journal of Food Science*, *68*(5), 1736–1742.

Charles, F., Sanchez, J., & Gontard, N., (2006). Absorption kinetics of oxygen and carbon dioxide scavengers as part of active modified atmosphere packaging. *Journal of Food Engineering*, *72*(1), 1–7.

Chauhan, O. P., Raju, P. S., Dasgupta, D. K., & Bawa, A. S., (2006). Modified atmosphere packaging of banana (cv. Pachbale) with ethylene, carbon dioxide and moisture scrubbers and effect on its ripening behavior. *American Journal of Food Technology*, *1*(2), 179–189.

Chinese Patent No. CN101601420A, (2009). *Controlled-Bactericide-Release Fruit and Vegetable Preservative Packaging Composite Film.*

Chinese Patent No. CN1040126A, (1988). *Germicide Capable of Preventing Apples from Putrefaction And its Use*.

Clarke, R., (2002). Intelligent packaging for safeguarding product quality. In: *Second International Conference on Active and Intelligent Packaging* (pp. 12–13). Gloucestershire, UK. Conference Proceedings. Campden & Chorleywood Food Research Association Group.

Coles, R., McDowell, D., & Kirwan, M. J., (2003). *Food Packaging Technology* (Vol. 5). CRC Press.

Cong, F., Zhang, Y., & Dong, W., (2007). Use of surface coatings with natamycin to improve the storability of Hami melon at ambient temperature. *Postharvest Biology and Technology*, *46*(1), 71–75.

Corrales, M., Fernandez, A., & Han, J. H., (2014). Antimicrobial packaging systems. *Innovations in Food Packaging*, 133–170.

Day, B. P., (2008). Active packaging of food. *Smart Packaging Technologies for Fast Moving Consumer Goods*. John Wiley, Hoboken, New Jersey.

Dijkink, B. H., Tomassen, M. M., Willemsen, J. H., & Van, D. W. G., (2004). Humidity control during bell pepper storage, using a hollow fiber membrane contactor system. *Postharvest Biology and Technology*, *32*(3), 311–320.

Dini, I., (2016). Use of essential oils in food packaging. In: *Essential Oils in Food Preservation, Flavor, and Safety* (pp. 139–147).

Esturk, O., Ayhan, Z., & Gokkurt, T., (2014). Production and application of active packaging film with ethylene adsorber to increase the shelf life of broccoli (*Brassica oleracea* L. var. Italica). *Packaging Technology and Science*, *27*(3), 179–191.

Fisher, K., & Phillips, C., (2008). Potential antimicrobial uses of essential oils in food: Is citrus the answer? *Trends in Food Science and Technology*, *19*(3), 156–164.

Gaikwad, K. K., Singh, S., & Lee, Y. S., (2017). A pyrogallol-coated modified LDPE film as an oxygen scavenging film for active packaging materials. *Progress in Organic Coatings*, *111*, 186–195.

Imran, M., Revol-Junelles, A. M., Martyn, A., Tehrany, E. A., Jacquot, M., Linder, M., & Desobry, S., (2010). Active food packaging evolution: Transformation from micro-to nanotechnology. *Critical Reviews in Food Science and Nutrition*, *50*(9), 799–821.

Jacobsson, A., Nielsen, T., & Sjöholm, I., (2004). Influence of temperature, modified atmosphere packaging, and heat treatment on aroma compounds in broccoli. *Journal of Agricultural and Food Chemistry*, *52*(6), 1607–1614.

Kerry, J. P., O'grady, M. N., & Hogan, S. A., (2006). Past, current and potential utilization of active and intelligent packaging systems for meat and muscle-based products: A review. *Meat Science*, *74*(1), 113–130.

Labuza, T. P., & Breene, W. M., (1989). Applications of "active packaging" for improvement of shelf life and nutritional quality of fresh and extended shelf-life foods 1. *Journal of Food Processing and Preservation*, *13*(1), 1–69.

Lanciotti, R., Gianotti, A., Patrignani, F., Belletti, N., Guerzoni, M. E., & Gardini, F., (2004). Use of natural aroma compounds to improve shelf life and safety of minimally processed fruits. *Trends in Food Science and Technology*, *15*(3/4), 201–208.

Lee, D. S., & Paik, H. D., (1997). Use of a pinhole to develop an active packaging system for Kimchi, a Korean fermented vegetable. *Packaging Technology and Science: An International Journal, 10*(1), 33–43.

Lee, D. S., (2016). Carbon dioxide absorbers for food packaging applications. *Trends in Food Science and Technology, 57*, 146–155.

Lee, D. S., Shin, D. H., Lee, D. U., Kim, J. C., & Cheigh, H. S., (2001). The use of physical carbon dioxide absorbents to control pressure buildup and volume expansion of Kimchi packages. *Journal of Food Engineering, 48*(2), 183–188.

Li, X., Li, W., Jiang, Y., Ding, Y., Yun, J., Tang, Y., & Zhang, P., (2011). Effect of nano-ZnO-coated active packaging on quality of fresh-cut 'Fuji' apple. *International Journal of Food Science and Technology, 46*(9), 1947–1955.

Lim, L. K., & Lee, K. E., (2010). U.S. Patent Application No.: 12/520, 294. Packaging system for the control of relative humidity of fresh fruits, vegetables, and flowers with simultaneous regulation of CO_2 and O_2. U.S. Patent and Trademark Office. Washington, DC.

Lord, J. B., (2008). The food industry in the United States. *Developing New Food Products for a Changing Market Place* (2nd edn., pp. 1–23). Boca Raton, Fla.: CRS Press.

Maneerat, C., & Hayata, Y., (2006a). Antifungal activity of TiO_2 photocatalysis against *Penicillium expansum in vitro* and in fruit tests. *International Journal of Food Microbiology, 107*(2), 99–103.

Maneerat, C., & Hayata, Y., (2006b). Efficiency of TiO_2 photocatalytic reaction on delay of fruit ripening and removal of off-flavors from the fruit storage atmosphere. *Transactions of the ASABE, 49*(3), 833–837.

Marbler, C. A., & Parmentier, R., (1999). *U.S. Patent No. 5,958,534*. Packaging material. U.S. Patent and Trademark Office, Washington, DC.

Mathlouthi, M., De Leiris, J. P., & Seuvre, A. M., (1994). Package coating with hydrosorbent products and the shelf life of cheeses. In: *Food Packaging and Preservation* (pp. 100–122). Springer, Boston, MA.

May, B. K., & Fickak, A., (2003). The efficacy of chlorinated water treatments in minimizing yeast and mold growth in fresh and semi-dried tomatoes. *Drying Technology, 21*(6), 1127–1135.

Mir, A. A., & Bandral, M. S. J. D., (2018). Effect of active packaging on quality and shelf life of peach fruits. *The Pharma Innovation Journal, 7*(4), 1076–1082.

Moreira, M. D. R., Pereda, M., Marcovich, N. E., & Roura, S. I., (2011). Antimicrobial effectiveness of bioactive packaging materials from edible chitosan and casein polymers: Assessment on carrot, cheese, and salami. *Journal of Food Science, 76*(1), M54–M63.

Mustonen, H. M., (1992). The efficacy of a range of sulfur dioxide generating pads against *Botrytis cinerea* infection and on out-turn quality of Calmeria table grapes. *Australian Journal of Experimental Agriculture, 32*(3), 389–393.

Ozdemir, M., & Floros, J. D., (2004). Active food packaging technologies. *Critical Reviews in Food Science and Nutrition, 44*(3), 185–193.

Patterson, B. D., & Joyce, D. C. (1993). A Package Allowing Cooling and preservation of Horticultural Produce Without Condensation or Dessicants. *International Patent Application PCT/AU93/00398*.

Polyakov, V. A., & Miltz, J., (2010). Modeling of the humidity effects on the oxygen absorption by iron-based scavengers. *Journal of Food Science, 75*(2), E91–E99.

Popplewell, L., & Henson, L., (2004). U.S. Patent Application No.: 10/202, 958. Packaging containing fragrance. U.S. Patent and Trademark Office. Washington, DC.

Rao, D. S., & Srivastava, A., (2018). Active and smart packaging techniques in vegetables. In Sudhakar, R. D. V., Anuradha, S., & Ranjitha, K., (eds.), *Advances in Postharvest Technologies of Vegetable Crops* (pp. 273–292). Apple Academic Press.

Realini, C. E., & Marcos, B., (2014). Active and intelligent packaging systems for a modern society. *Meat Science, 98*(3), 404–419.

Rico, D., Martin-Diana, A. B., Barat, J. M., & Barry-Ryan, C., (2007). Extending and measuring the quality of fresh-cut fruit and vegetables: A review. *Trends in Food Science and Technology, 18*(7), 373–386.

Robertson, G. L., (2013). *Food Packaging: Principles and Practice* (3rd edn.). CRC Press, Boca Raton, USA.

Rojas-Grau, M. A., Soliva-Fortuny, R., & Martin-Belloso, O., (2009). Edible coatings to incorporate active ingredients to fresh-cut fruits: A review. *Trends in Food Science and Technology, 20*(10), 438–447.

Rooney, M. L., (1995). *Active Food Packaging*. Chapman & Hall, London, UK.

Rux, G., Mahajan, P. V., Geyer, M., Linke, M., Pant, A., Saengerlaub, S., & Caleb, O. J., (2015). Application of humidity-regulating tray for packaging of mushrooms. *Postharvest Biology and Technology, 108*, 102–110.

Scully, A. D., & Horsham, M. A., (2007). *Active Packaging for Fruits and Vegetables* (pp. 57–73). CRC Press, Boca Raton, Florida.

Serrano, M., Martinez-Romero, D., Castillo, S., Guillen, F., & Valero, D., (2005). The use of natural antifungal compounds improves the beneficial effect of MAP in sweet cherry storage. *Innovative Food Science and Emerging Technologies, 6*(1), 115–123.

Serrano, M., Martinez-Romero, D., Guillen, F., Valverde, J. M., Zapata, P. J., Castillo, S., & Valero, D., (2008). The addition of essential oils to MAP as a tool to maintain the overall quality of fruits. *Trends in Food Science and Technology, 19*(9), 464–471.

Sezer, E., Ayhan, Z., Çellkkol, T., & Guner, F., (2017). Effect of zeolite added active packaging material on the quality and shelf life of kiwifruit. *GIDA-Journal of Food, 42*(3), 277–286.

Sharma, A., & Thakur, N. S., (2016). Influence of active packaging on quality attributes of dried wild pomegranate (*Punica granatum* L.) arils during storage. *Journal of Applied and Natural Science, 8*(1), 398–404.

Silveira, A. C., Moreira, G. C., Artes, F., & Aguayo, E., (2015). Vanillin and cinnamic acid in aqueous solutions or inactive modified packaging preserve the quality of fresh-cut cantaloupe melon. *Scientia Horticulturae, 192*, 271–278.

Singh, R., Giri, S. K., & Kotwaliwale, N., (2014). Shelf-life enhancement of green bell pepper (*Capsicum annuum* L.) under active modified atmosphere storage. *Food Packaging and Shelf Life, 1*(2), 101–112.

Sun, X., Bai, J., Ference, C., Wang, Z., Zhang, Y., Narciso, J., & Zhou, K., (2014). Antimicrobial activity of controlled-release chlorine dioxide gas on fresh blueberries. *Journal of Food Protection, 77*(7), 1127–1132.

Suppakul, P., Miltz, J., Sonneveld, K., & Bigger, S. W., (2003). Active packaging technologies with an emphasis on antimicrobial packaging and its applications. *Journal of Food Science, 68*(2), 408–420.

Tarr, C. R., & Clingeleffer, P. R., (2005). Use of an oxygen absorber for disinfestation of consumer packages of dried vine fruit and its effect on fruit color. *Journal of Stored Products Research, 41*(1), 77–89.

Vermeiren, L., Devlieghere, F., Van, B. M., De Kruijf, N., & Debevere, J., (1999). Developments in the active packaging of foods. *Trends in Food Science and Technology*, *10*(3), 77–86.

Warad, H. C., & Dutta, J., (2005). Nanotechnology for agriculture and food systems: A review. *Proceedings of the Second International Conference on Innovations in Food Processing Technology and Engineering*. Bangkok. Pathumthani, Thailand: Asian Institute of Technology.

Wilson, C. L., (2007). *Intelligent and Active Packaging for Fruits and Vegetables*. CRC Press.

Yam, K. L., Takhistov, P. T., & Miltz, J., (2005). Intelligent packaging: Concepts and applications. *Journal of Food Science*, *70*(1), 1–10.

CHAPTER 7

MODIFIED ATMOSPHERE PACKAGING FOR HORTICULTURAL PRODUCTS

TANWEER ALAM,[1] PRITI KHEMARIYA,[2] and MOHAMMED WASIM SIDDIQUI[3]

[1]*Indian Institute of Packaging, Mumbai, Maharashtra, India, E-mail: amtanweer@rediffmail.com*

[2]*Indian Institute of Packaging, Delhi, India*

[3]*Bihar Agricultural University, Sabour, Bihar, India*

ABSTRACT

Modified atmospheric packaging (MAP) is a one of the important packaging system which enhances the shelf life of horticultural produces. This system includes one time alteration in gas composition inside the storage chamber by incorporating either the single or the combination of the gases. Nitrogen is frequently used to reduce the concentration of oxygen and carbon dioxide in the package. In the present chapter, the role of gases used, the type of packaging materials used, indicators and factors affecting the MAP has been discussed along with recent advances done in the field of MAP for packaging of various horticultural produces.

7.1 INTRODUCTION

Indian agriculture contributes 14.5% GDP to the economy. It sustains 60% population of our country. In the agriculture, horticulture sector plays an important role not only to provide us nutritional and healthy foods, but also generate a cash income to growers. It shares 33% output in Indian agriculture. India is quiet able to grow almost all F&V but these are grown on only 7–8% of gross cropped area. In the year the 2012–13, as per the report given in Horticulture-Statistical Year Book India 2017, the productivity for F&V has been 11.6 and 17.6 tons/ha, respectively. Moreover, there is a great percentage (20–40%) of total postharvest losses due to very less (3.5%) processing reported every year of F&V. These percentages are not acceptable and adversely affect the Indian economy. Instead, 32% share of the total food market is contributed by the Indian food processing industry. As per the Economic survey 2016–17, the following data has been reported for the contribution given by the Indian food processing industry:

Gross value added (GVA) in manufacturing	8.8
Gross value added (GVA) in agriculture	8.39
India's exports	13%
Total industrial investment	6%

Postharvest losses for horticultural produce are, however, difficult to measure. Some vegetables in India have been estimated 100% postharvest losses. There are many reasons for a great percentage of postharvest losses of F&V. Handling of raw F&V is carried out through many stages of middlemen. Management and processing of raw produce is very low or negligible. In rural areas, entrepreneurial urge are also very low. The major drawback of this sector is the unavailability of cold chain and machinery facilities. However, it has been reported that sufficient investment has been made for cold chain establishment by the National Centre for Cold Chain Development (Hegazy, 2013).

7.2 POST HARVEST PROCEDURE

It is a systematic stepwise procedure from harvesting of horticultural products to processing to packaging to transportation and distribution from growers level to consumers. The flow diagram of the stepwise procedure of postharvest processing is shown as flow diagram Figure 7.1.

7.3 PACKAGING OF HORTICULTURAL PRODUCTS

Packaging is one of the more important steps of postharvest processing to supply horticultural products from growers to consumers. According to recent FAO, 50% of agricultural products are destroyed without packaging.

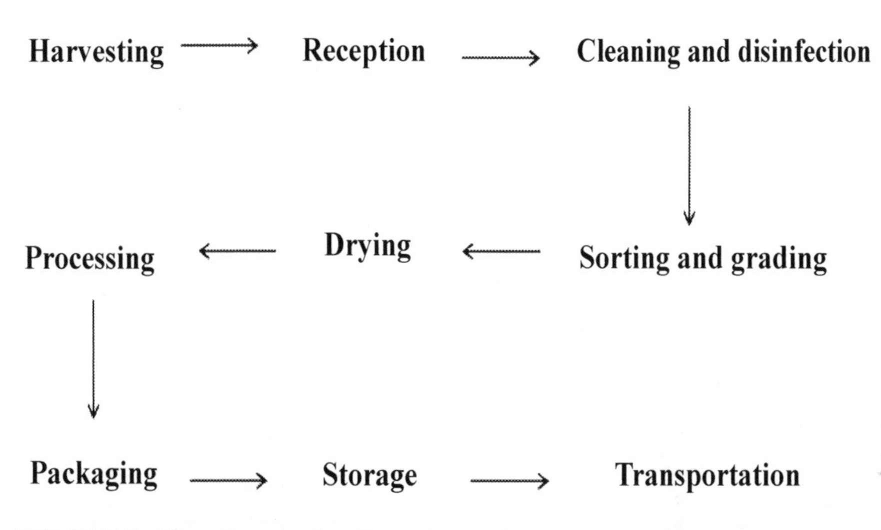

FIGURE 7.1 Flow diagram of postharvest procedure.

Packaging plays an important role in the fruit and vegetable distribution chain. It plays four main functions first is containment for movement of products, second protection and preservation against environment factors (dust, water, etc.), handling, and transportation damages, third convenience for handlers and consumers and forth communication for detailing information (type, source, gross weight, net weight and unit size) required by government regulations (Watkins and Nock, 2012). Several types of packaging materials are used for produce around the world. These are available in a variety of materials as well, such as plastic, corrugated fiberboard (CFB), wood, and biodegradable bioplastics and fibers. The plastic packages are generally manufactured from polyethylene terephthalate (PET). These are necessary for certain commodities; however, corrugated, and non-CFB is used to pack fresh horticultural products. Wooden containers such as wire-bound are a traditional form of packages is also used to pack fresh produce. Although, wooden packages have limitations as these are relatively heavy, expensive, and abrasive to the F&V with great disposal issues (http://edis.ifas.ufl.edu).

7.4 TYPES OF PACKAGING

There are different types of packaging systems used to preserve and protect horticultural products such as passive packaging, active packaging, smart packaging, modified atmosphere packaging (MAP), intelligent packaging, flexible packaging, antimicrobial packaging, vacuum packaging (VP), moderate VP and aseptic packaging, etc. In the present chapter, Modified Atmospheric Packaging (MAP) to enhance shelf life of horticultural produces is described in detail.

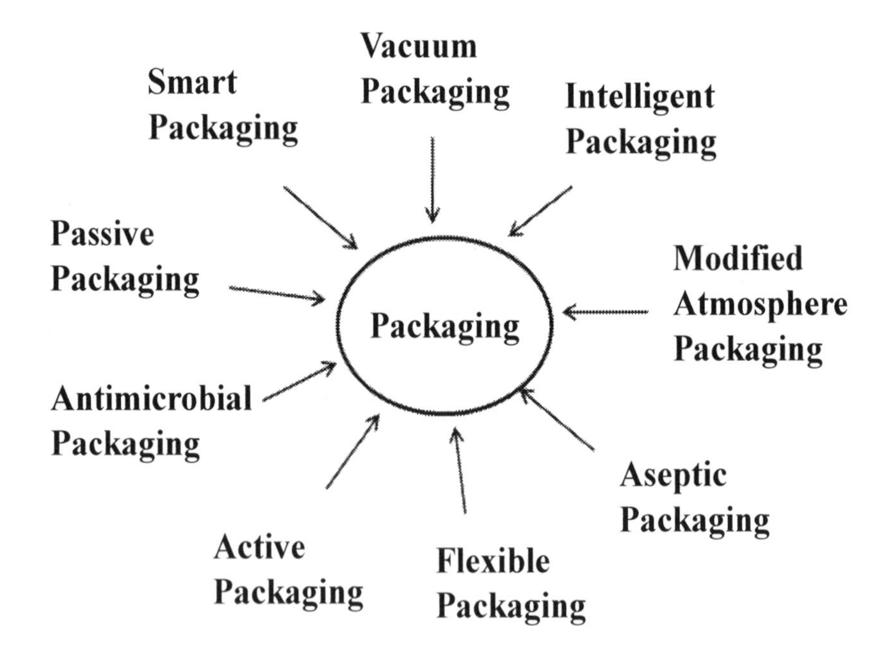

7.5 MODIFIED ATMOSPHERIC PACKAGING (MAP)

The consumer's demand for freshly packaged food products has been increasing enormously. The concept of MAP is not recent; however, the demand for MAP and controlled atmosphere (CA) packaged (CAP) products is increasing in the present time. MAP is a preservation technique of packaging of postharvest highly respiring fresh produces under CA used to retain the quality of fresh food produce, to extend the shelf life of fresh and minimal processed food items during storage period and to reduce the waste loss percentage. MAP retains high organoleptic quality of fresh food products by controlling two factors, namely respiration of food products and production of ethylene by food products. The exact working mechanism of MAP is the reduction of oxygen and/or increase of carbon dioxide concentration inside polymeric film packages. During the process operation, one time alteration in gas composition is performed inside the storage chamber by incorporating either the single or the combination of the gases. Nitrogen is frequently used to reduce the concentration of oxygen and carbon dioxide in the package. There are possibilities in alteration of modified gaseous composition of storage chamber due to the following reasons:

1. Diffusion of gases from outside or inside the produce;
2. Permeation of gases from outside or inside the package;
3. Microbial activity inside the storage chamber.

The gaseous atmosphere is maintained by vacuum, gas flushing, and by controlled permeability by using specific barrier packaging films. Barrier packaging films reduces excess moisture loss, diffusion of gases, and incidence of contamination inside the food package. Continuous gas stream is flushed in food package to replace the air in gas flushing technique. By this procedure, 2–5% oxygen concentration is maintained inside the food package by incorporation of nitrogen gas. This process decreases the oxidation process inside the food package and thereby reducing discoloration, spoilage, and off-flavors. In another method, i.e., compensated vacuum technique involves the creation of vacuum first inside the food package followed by desired gaseous composition is maintained.

Initially, MAP was used for enhancing the shelf life of apple in the year 1927. Commercially it came in the market in the early 1970s in Europe. Nowadays, this technique is widely used for shelf-life extension of wide range of products including fresh and chilled foods, fruits, and vegetable-based processed products, fresh fish, seafood, raw, and cooked meat, red meat, poultry, beverages, ready meals, dried foods, and bakery products.

The main objective of MAP is to reduce the microbial growth rate so as to make the item organoleptically acceptable and health-wise safe. In F&V the commonly found microorganisms belong to *Pseudomonas, Erwinia, Xanthomonas, Enterobacter, Flavobacterium, Leuconostoc, Lactobacillus*, yeasts, and molds. Pathogenic organisms include *Clostridium botulinum, Yersinia enterocolitica, Listeria monocytogenes, Aeromonas hydrophila, Bacillus cereus,* and *Campylobacter jejuni*. MAP delays senescence of F&V and thus makes them less susceptible to pathogens (Figure 7.2).

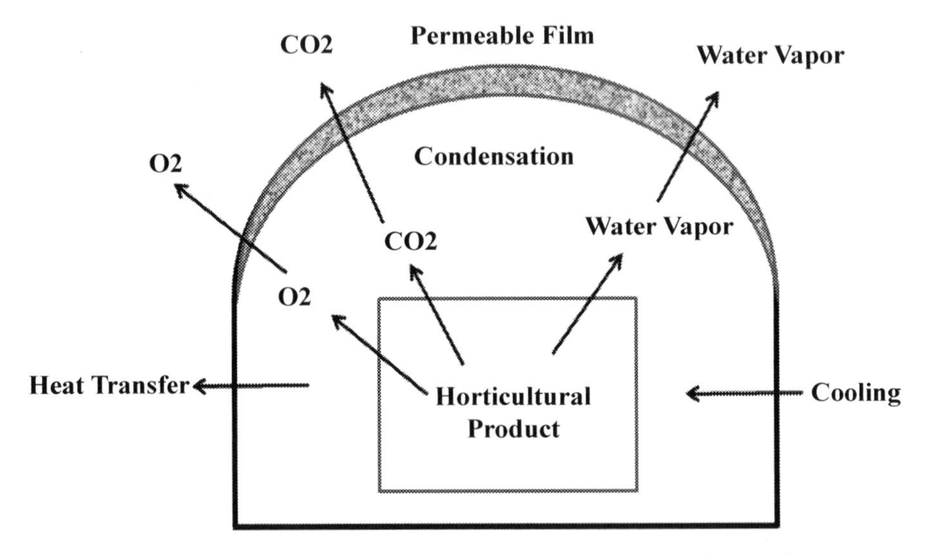

FIGURE 7.2 Mechanism of action in modified atmosphere packaged horticultural product.

7.6 ROLE OF GASES USED IN MAP

The commonly used gases in MAP are Oxygen, CO_2, and N_2 however; several studies are being carried out by scientist to observe the efficacy of other gases such as nitric oxide, nitrous oxide, ethylene, sulfur dioxide, propylene oxide, chlorine, ozone, etc. However, their commercial application is still under consideration as for as their cost-effectiveness and safety regulations are concerned. Generally, 1–5% oxygen gases are recommended for MAP of horticultural products while maintaining their quality and shelf life. This reduction of oxygen concentration from 20% to up to 0% is performed for slowing down the aerobes growth and for speed up oxidation reactions. The reduced oxygen volume inside the package is replaced by nitrogen or carbon dioxide. Nitrogen is an inert gas and does not perform any preservation process, while carbon dioxide inhibits microbial growth by lowering the pH of the food package. This gas also maintains the red coloration of meat. In the MAP, three types of gas, i.e., O_2, CO_2, and N_2 and following three kinds of gas mixtures are used:

1. Nitrogen as Inert packaging;
2. Semi-reactive blanket (CO_2/N_2 or $O_2/CO_2/N_2$); and
3. Fully reactive blanket (CO_2 or CO_2/O_2).

7.6.1 CARBON DIOXIDE

Carbon dioxide is colorless with little pungent odor having bacteriostatic and fungistatic effect. The gas (20%) selectively inhibits the growth of gram-negative bacteria, such as pseudomonas. Lactic acid bacteria (LAB), such as streptococci and lactobacilli, are less affected by elevated levels of Carbon dioxide. It alters the nutrient uptake and absorption function of the cell membrane of spoilage organism due to its ability to penetrate the cellular membrane, which causes intracellular pH changes and inhibits enzyme action or decreases the rate of enzymatic reactions hence disrupts internal enzymatic equilibrium. Carbon dioxide turns into acidic carbonic acid after reacting with water leading and thus reduces the pH of the most products. CO_2 during log growth phase exhibited antimicrobial effect and because of this activity lag phase remains for long period. Application of CO_2 should be done in balanced amount otherwise tainting of flavor, water loss, and collapsing of package may happen. Therefore, a balance should be maintained so that product with increased shelf life and without any negative effect should reach to customer.

7.6.2 OXYGEN

Oxygen is a prime requirement for aerobes. Oxygen is colorless, odorless, and very reactive gas which promotes the deteriorative reactions of foods. Due to the availability of oxygen inside the package, oxidation of fat content, oxidation of pigments and browning in the products could be happened. In MAP, two processes are followed:

1. Removal of oxygen to inhibit the growth of spoilage microorganisms inside the package.

2. Maintaining the low levels of oxygen inside the package. This is done for the following reasons:
   - To maintain fresh, natural color of commodity;
   - To reduce food deterioration;
   - To maintain respiration of fruit and vegetables;
   - To inhibit the microbial growth.

7.6.3 NITROGEN

Nitrogen is used to replace the oxygen content inside the food package. It does not react with the food as it is an inert gas. It acts as filler gas (balance gas) without odor and taste and keeps the package flexible by developing a vacuum. It acts as a cushion and therefore prevents the collapsing of the package. Argon is also used to flush out air from the package. Argon delays metabolic rate of food of the package due to its interference with oxygen receptor site on enzyme. Because of its high cost, nitrogen is generally recommended (Table 7.1).

TABLE 7.1 Optimum Percentage Gases (CO_2 and O_2) for Fruits and Vegetables

Product		**Percentage of CO_2**	**Percentage of O_2**
Fruits	Strawberry	Up to 20	Up to 10
	Grape	Up to 10	Up to 10
	Papaya	Less than 8	Less than 5
	Mango	Less than 8	Less than 7
	Orange	Less than 5	Up to 10
	Kiwi	Less than 5	Less than 2
	Banana	Less than 2	Less than 3
Vegetables	Cauliflower/Broccoli	3–4	2–3
	Cucumber	Nil	Less than 5
	Capsicum	2	Less than 5
	Tomato (ripe)	Up to 10	Less than 5
	Bean sprouts	Up to 15	5
	Lettuce	Nil	1–3
	Sweet corn	2	Up to 10

Source: Guide, Packaging Fresh Fruits and Vegetables, Danish Technological Institute, Packaging, and Transport (2008); modified by authors.

7.7 PACKAGING MATERIALS USED FOR MAP

The packaging materials are the main feature for modified atmosphere. It possesses permeability for different gases. There are many types of plastic materials with different properties. The selection of correct packaging material is very essential to prevent the loss of package atmosphere. The packaging material should be flexible, heat sealable, relatively transparent, recyclable, and strong having resistance to puncture, chemical, and degradation, antifogging properties, water transmission rate and permeability for carbon dioxide and oxygen. It should be low cost, non-toxic, chemically inert, soft, durable, and lightweight. The packaging materials used in MAP provides the following benefits:

- It reduces contamination in the food packages during handling;
- It reduces the exposure of light in the food packages;
- It maintains high relative humidity in the food packages;
- It reduces water loss from the commodity;
- It creates barriers to spread spoilage from one to other commodities;
- It facilitates brand identification.

The modified atmosphere multilayer packaging is also applicable in food packaging. Because many times it is difficult for a single packaging material to have all the characteristics of protection, technical, and other necessity for packaging in modified atmosphere. To achieve the best properties of polymeric films, the plastic packaging films are combined with other packaging materials such as paper or aluminum. This combination process is carried out by coating, lamination, co-extrusion, and metallization processes.

The packaging materials used for MAP are made up of basic polymers, i.e., polyvinyl chloride (PVC), polyethylene (LDPE, HDPE, MDPE), polypropylene (BOPP), ionomers, polycarbonate films, polybutylene polyester (PET/PEN), polyolefin, polystyrene, poly chloro trifluoro ethylene (PCTFE), ethylene vinyl alcohol, polyamide (nylon-6), polyvinylidene chloride (PVDC), ethylene-vinyl alcohol (EVOH), cellulose-based plastics, biodegradable polymers and polyethylene terephthalate.

Polyethylene, a hydrocarbon polymer is one of the important packaging materials. Commercial it is produced with a variable amount of branching within the polymer. HDPE has the minimum branching. Therefore, HDPE possesses the greatest thermal stability and lowest permeability. HDPE is flexible, lightweight, cheap, strong, and tough but poor clarity having better barrier properties, resistance to chemicals and moisture, permeable to gases. LDPE has a comparatively high ratio of CO_2 and O_2 permeability. LDPE is soft, lightweight, recyclable, low cost, heat sealable, relatively transparent and flexible has good tensile strength, burst strength, impact resistance, barrier to water vapor and tear strength, retaining its strength up to –60°C. LDPE is not a good barrier to gases. It is suitable for F&V having low respiration rate, e.g., tomato, apple, carrot, etc. It is not suitable for strawberry as having high respiration rate. LLDPE is soft, inert, flexible, strong material having better impact strength, tear resistance, tensile strength, and elongation, resistance to grease, moisture, environmental stress cracking, and puncture. It is not suitable for applications which require significant exposure to heat. Polypropylene (BOPP) is stronger, denser transparent and high gas and moisture vapor barrier than the polyethylene. It possesses excellent grease resistance. It's strength, clarity, and durability is more than the LDPE. Polyesters (PET/PEN) has excellent transparency and mechanical properties, good resistance to heat, mineral oil, solvents, and acids chemical degradation, barrier to gases and moisture, odors, and flavors. It is recyclable, glass-like transparent having higher cost among plastics. Polyvinyl chloride has efficient permeability for gas and water vapors. It is strong, recyclable, and transparent with high clarity, excellent resistance to chemicals, oils/fats and grease, etc. PVDC is heat sealable. It is having high barrier properties to gases and water vapor. It is used in hot filling, retorting, low-temperature storage, etc. Polystyrene is chemically inert with a high degree of clarity, gas permeability, tensile strength, transparency, but poor barrier to moisture vapor and gases. Polyamide (Nylon-6) is strong with good oxygen barrier, odor, and flavor barrier, chemical resistance and high-temperature performance.

7.8 ADVANCES IN PACKAGING MATERIALS OF MAP

Based on the respiration rate of the food packed inside the package, different types of packaging materials are used in MAP. In spite of commonly used polymers, various R and D are also developing innovative packaging materials with improved property and functionality for MAP, including biodegradable films, antioxidant active films, micro-perforated films, and nano-active films.

Antioxidant Active films are active packaging films with antioxidants properties and thereby preserve the food by keeping nutritional and organoleptic quality. This type of Packaging film is developed by incorporating antioxidant compounds with packaging materials. This incorporation reduces the steps of spraying, mixing, or immersion of antioxidants with food. This film possesses low active substance concentration that allows slow incorporation of antioxidants in the food. Nano-active films is made up by incorporating some nanosized metals (Siver, Titanium oxide, etc.), with the packaging films, which results in the improvement in functionality of film such as antimicrobial property to prevent microbial growth. Nano metals disrupt the permeability of microbial cell membrane, enzyme system, and metabolism. In order to maintain adequate respiration rate of food items inside the package, the permeability of packaging materials for oxygen and carbon dioxide is enhanced by using micro-perforated packaging films.

The film has a ratio of O_2/CO_2 transmission nearing 1, which means the headspace area contains the gaseous composition at its higher concentration (Table 7.2).

TABLE 7.2 $O_2:CO_2$ Ratio for Packaging Materials

SL. No.	Packaging Material	$O_2:CO_2$ Ratio
1.	Polyvinyl chloride	3–7
2.	Polyester	3–3.5
3.	Polypropylene	3–6
4.	Polystyrene	3–4
5.	LDPE	2–6

Source: Kader (2002); modified by authors.

7.8.1 EXAMPLES OF MAP

- **Gas Flushing:** Nitrogen gas is applied to flush out the oxygen present within the package to prevent the spoilage by delaying oxidation and decreasing the growth of aerobes. It also acts as cushioning the package against any injury.
- **Barrier Packaging Films:** Specific packaging films as a barrier to providing increased protection is used for MAP. Barrier film decreases permeability to moisture and oxygen. Examples of barrier films are low-density polyethylene (LDPE), polyvinylchloride (PVC), polypropylene (PP), etc. Nowadays, smart packaging films with temperature indicator, leakage indicator, and edibility indicator are widely used.
- **Scavenger or Desiccant Sachet:** Oxygen scavengers or sachet filled with desiccant are also used in MAP to absorb ambient moisture and oxygen. The sachet contains catalysts or activators. The combination

of powdered iron with ascorbic acid and/or activated carbon is one of the common examples of this type of packaging film.

- **On-Package One Way Valves:** These valves in the MAP machine permit some gases to diffuse out from the food package. These valves don't allow the entry of any gas from outside.

7.9 TYPES OF MAP

7.9.1 PASSIVE MAP

Passive modified packaging is done by modifying the package atmosphere as per the rate of O_2 consumption and CO_2 evolution from the food items inside the packaging film and permeability of package. This leads to the atmosphere with more CO_2 and less O_2 concentration. After a certain time period, the gas composition of CO_2 and O_2 in the product package reaches a definite equilibrium, i.e., the total amounts of CO_2 produced and O_2 consumed become the same. The gas composition at the equilibrium state is determined by horticultural products, environmental factors, and package permeability. Because each product exhibits distinct behavior so that passive MAP has to be standardized for each commodity.

7.9.2 ACTIVE MAP

In this technique, a slight vacuum is pulled inside the storage chamber which later on replaced by the gaseous composition of CO_2, O_2, and N_2 gases as per the required rate of respiration inside the packaging film. In this system, the gaseous composition inside the package remains balanced because consumption of oxygen and evolution of carbon dioxide by/from food items is balanced by the oxygen and carbon dioxide supplied from outside. Active MAP is quicker than Passive MAP as the equilibrium is maintained rapidly in Active MAP. Different types of absorbers such as oxygen scavenger, carbon dioxide scavengers, ethanol emitters, and ethylene scavenger are also used with Active MAP to maintain oxygen and carbon dioxide levels inside the package. This step enhances the shelf life of the package by modifying the gaseous composition of headspace. The absorbers which are used in active MAP, must be harmless for human beings, should have efficient storage stability and large capacity of gas absorption.

7.10 INDICATORS USED IN MAP

Following indicators are generally used with MAP:

1. Time-temperature indicators to determine the remaining storage period of commodities;
2. O_2 indicator;
3. CO_2 indicator;
4. Pathogen growth indicator.

7.11 FACTORS AFFECTING MODIFIED ATMOSPHERE PACKAGING (MAP)

The composition of the atmosphere inside the packaging results from the interaction of several following factors:

7.11.1 CONTROLLABLE FACTORS

1. Permeability characteristics of packaging materials;
2. Respiration rate of atmosphere and commodity;
3. Types of selected packaging materials suitable for specific permeability;
4. Concentration of gases within the package;
5. Storage temperature.

7.11.2 UNCONTROLLABLE FACTORS

1. Cultivar of perishable horticultural commodity;
2. Cultural practices for horticultural products;
3. Developmental stage of horticultural products;
4. Harvesting management;
5. Postharvest management of produce:
   - In MAP, relative humidity (RH) is a very important factor because optimum RH reduces the rate of transpiration of food items resulting in the reduced rate of wilting and firmness degradation. Therefore, in almost all sealed packages contains RH at saturation level.

- Various factors such as water vapor transmission rate (WVTR) of packaging materials, fluctuated temperature of surrounding environment, moisture release from food items inside package increases vapor content or water condensation inside the food package. This leads to the development of unfavorable package atmosphere such as pathogens proliferation, decay, regrowth, etc.

7.12 QUALITY ASSURANCE IN MAP PACKAGES

MAP is a well-established preservation technique to maintain the quality of package. However, the accuracy in maintaining the in-package gaseous atmosphere has essential to maintain otherwise product spoilage may occur. Oxygen level, gas levels in cylinder, condition of sealing bars should be proper to prevent poor package and can cause spoilage. Similarly, headspace gas analyzer, on-line gas analyzers, and leak detectors should be checked to work properly in order to ensure enhanced self-life. Compatibility of packaging films with fresh produce and in-package temperature management is also a critical factor. Because incompatible film and high temperature may cause accumulation of compounds such as CH_3CHO, C_2H_5OH, $CH_3COOC_2H_5$, lactic acid, etc., due to anaerobic respiration this can cause off-flavor and taste leading to deterioration of fresh produce. Some other parameters such as physiological disorder, fruit ripening, and decay susceptibility are also very important for MAP.

7.13 RECENT ADVANCES ON PACKAGING OF HORTICULTURAL PRODUCTS WITH MODIFIED ATMOSPHERE

Several studies on the packaging of horticultural products with modified atmosphere have been done by various researchers in the recent time. The influence of modified atmospheric pressure on the quality of baby spinach was studied for the treatments consisted of control [(normal air) (78% N_2; 21% O_2)], modified atmosphere (MA) (5% O_2; 15% CO_2; balance N_2) at storage temperature (4, 10, and 20°C for 0, 3, 6, 9, and 12 days of storage. The overall reduction in O_2 and increase in CO_2 levels was found in the headspace gas over the storage period. The total antioxidant activity and flavonoids were well maintained at 4°C for 9 days of storage (Mudau et al., 2018).

The applicability of MAP for improving the food quality and safety of various fresh produce (grape, soybean sprout, lotus root, green vegetables) was studied during refrigeration storage (4 ± 2°C) in the presence of air and 100% CO_2 gas. In the 100% CO_2 gas, Low count of total mesophilic bacteria, *Escherichia coli*/coliform, and yeast/mold was found (Hyun and Lee, 2017).

Saffron was tested for 3 months period for observing moisture content and sensory attributes under MAP. During the experiments, different gaseous combinations were prepared to be kept on different temperature such as 35, 25, and 4°C. The gaseous combination of CO_2 and N_2 was as follows:

- MAP1 (100% N_2);
- MAP2 (40% N_2 + 60% CO_2);
- MAP3 (60% N_2 + 40% CO_2).

At every three weeks interval, all samples were analyzed and was significant with MAP1 than rest two MAPs.

A study was conducted in MAP with diced onion to observe the effect of sanitizers, in-package condition, and their interactions on physic-chemical attributes and the growth profile of selected microorganisms such as *Salmonella typhimurium*, mesophilic aerobic bacteria, yeast, and mold, for the storage period of 14 days at 7°C. The sanitizers selected for the study were sodium hypochlorite, peroxyacetic acid, or liquid chlorine dioxide. It has been concluded that Sodium hypochlorite and elevated CO_2/reduced O_2 was the best sanitizer and gaseous combination for shelf life extension of in-packaged diced onions (Page et al., 2016).

Two fresh-cut vegetable swede (*Brassica napus* L. var. napobrassica (L) Rchb.) and turnip (*Brassica rapa* L. ssp. rapiferaMetzg.) were tested for Passive and active MAP. Active MAP was done by flushing 5% O_2 in a sealed package. The test was conducted at 5 and 10°C for 5 and 10 days. Significantly decreasing trend was obtained for glucose, fructose, and sucrose, and the increasing trend was noticed with total aliphatic and indolic glucosinolates in swede during storage. Turnip has shown decreased total aliphatic glucosinolates and an increased rate of total indolic glucosinolates. It was observed that changes in color, appearance, flavor, and taste can be prevented by maintaining low temperature and less storage period (Helland et al., 2016).

European plums cultivars ('Ramasin' and 'Ariddo di Core') were studied with MAP for 21 days of storage at 1 ± 1°C with 90–95% RH by Giuggioli et al. (2016). In this study, 90 and 65 μm sized films have shown offered better effectiveness in both cultivars. This study has exhibited the applicability of integrating multiple indexes in the analysis of shelf life procedures.

Quality attributes of fully ripe fruits of strawberries cultivar 'Senga Sengana' from organic and 'Arosa' from conventional production was investigated under MAP at 1°C for 5 days under different conditions and compared with the control sample. 'Arosa' fruits were firmer if MAP-treated, irrespective of pre-cooling and exhibited minimum spoilage in MAP with pre-cooled strawberries. 'Senga Sengana' showed the firmness and was not affected by the treatments. Chemical properties of treated samples were not affected as well. Hence, it has been concluded that MAP extends the shelf life of fully ripe strawberries.

The efficiency of different plastic films (pp, LDPE, and HDPE) was evaluated for Eva apples during cold storage at 0.5 ± 0.5°C for 225 days. Respiration, ethylene production, firmness, mass loss, total pectin, soluble pectin, soluble solids, total acidity, and epidermis background color of each treatment group were evaluated at 45, 135, and 225 days during storage. The HDPE has not exhibited a decrease in ethylene production during storage and the fruits were with greater firmness. Smaller percentage of mass loss was observed during the storage. Moreover, the Eva apple cultivar under modified atmosphere can be stored for up to seven months (Fante et al., 2014).

A study on shredded white cabbage was studied at 0°C with MAP in which 1.5% O_2 and 17% CO_2 concentration retained color, exterior appearance and organoleptic attributes. It has reduced the growth of pathogens causing pepper spot' disorder (Manolopoulou and Varzakas, 2013).

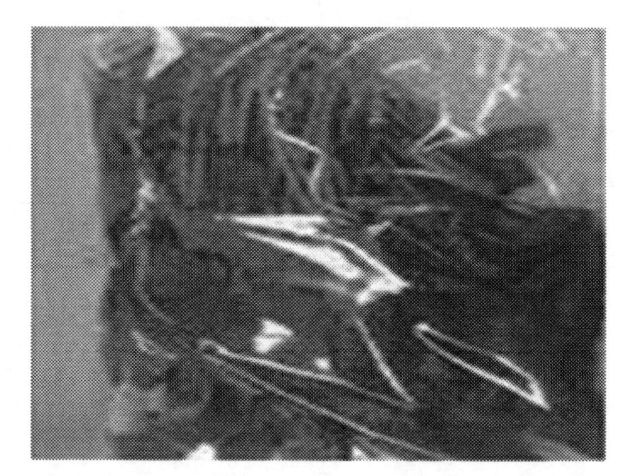

The proportions of O_2, CO_2, and N_2 in MAP were studied on spinach at room temperature by measuring weight loss, respiration speed, Vitamin C content, chlorophyll content, and sense index. 10% O_2 + 10% CO_2 + 80% N_2 gaseous composition has showed better preservative efficiency and enhanced shelf life by 8 days (Yang et al., 2012).

A mathematical model was formulated with active and passive MAP to predict gas changes of Endives packed in LDPE pouches at 5C and 20°C with and without individual oxygen or carbon dioxide scavenger sachet. Respiration rate in endives, transmission rate of gases through LDPE, absorption kinetics of gas scavengers, and temperature were measured. Oxygen scavengers reduced by half the transient period duration (50 h compared with 100 h without scavengers) without modifying the gas equilibrium composition. The results obtained suggested a beneficial influence on keeping the quality of the product (Charles et al., 2005).

Active MAP in LDPE/commercial iron-based scavenger system for tomatoes was performed at 20°C by formulating a mathematical model to predict characterization of absorber (Charles et al., 2003).

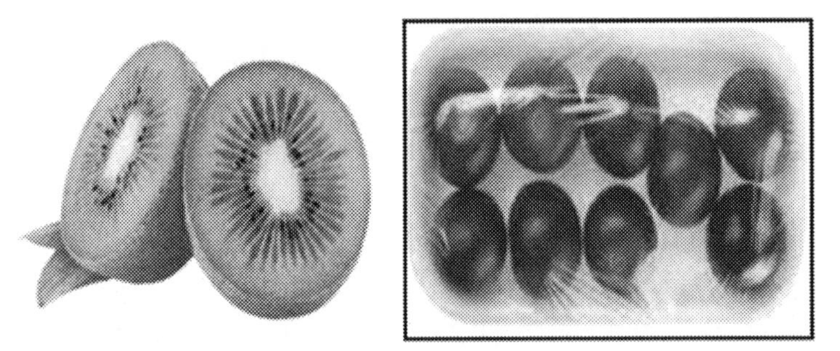

The modified atmosphere was developed in the wrapped trays at 20 and 10°C to prolong the shelf life of freshly harvested 'Hayward' kiwifruit. Low ethylene concentrations and favorable gaseous composition has enhanced the shelf life of kiwifruit with normal ripening and delayed softening at 20 and 10°C (Kitsiou and Sfakiotakis, 2003).

7.14 ADVANTAGES AND DISADVANTAGES OF MODIFIED ATMOSPHERE PACKAGING (MAP)

7.14.1 ADVANTAGES OF MAP

It improves presentation, clear view and all-round visibility of package. MAP reduces the handling and distribution of undesired and low-grade commodities and labor and waste at the retail level. MAP provides safety to sealed products from leakage and odor loss. MAP provides products to consumers at their desired demand. It is the preservation technique without chemical or physical preservatives.

It is beneficial practice over conventional methods of preservation. In MAP, little or no chemical preservatives are used. With the MAP, extension of product's shelf life can be achieved in the presence of less oxygen content and thus no aerobes growth and also provides preservative activity before the growth of anaerobes. Successful control of product respiration rate, ethylene production, yellowing, and browning decay maintains the highest sensory score by MAP. It also reduces the sensitivity to ethylene and thus delays the ripening of fruits and vegetable produces. The gaseous modification arrests the respiration and senescence processes and thus prolong the shelf life of perishable commodities of the package. With the MAP, the shelf life of meat can be enhanced up to three weeks. Similarly, Cheese shelf life can be achieved up to six months. It retains the chlorophyll content and moisture content of F&V. This packaging system decreases the incidence and severity of disease-causing microorganisms inside the storage chamber as well.

7.14.2 DISADVANTAGES OF MAP

The MAP machinery, including gas-packaging machinery, gases, and packaging materials, analytical equipment to ensure correct gas mixtures are very costly and not easily affordable. Moreover, there is the possibility of growth of microorganism in the suitable temperature and thus can cause foodborne illness and improper packaging by retailers and consumers. The non-biodegradable plastic films used for packaging is environmentally harmful.

7.15 CONCLUSION

MAP is a multidisciplinary technology in which synthetic chemicals are not used and no residues are remained hence no major environmental issues are generated. With the time many advances have been performed with MAP such as improved gas diffusion characteristics from the polymeric films and increased microenvironment by availability of various absorbers (O_2, CO_2, water vapors and ethylene). However, the microbiological safety aspect has not been widely studied with respect to MAP of horticultural products. Therefore, with other technological future advancements the microbiological safety studies is also needed. Moreover, different fresh products require specific MAP parameters, and also no any universal criteria are available to follow for all fresh horticulture products. Therefore, the innovative studies should be done to enhance the technical criteria for all commercially available fresh products to extend the shelf life.

KEYWORDS

- **controlled atmosphere packaged**
- **ethylene-vinyl alcohol**
- **gross value added**
- **low-density polyethylene**
- **modified atmosphere packaging**
- **poly chloro trifluoro ethylene**
- **polyethylene terephthalate**
- **polypropylene**

REFERENCES

Charles, F., Marcano, J. S., & Gontard, N., (2003). Active modified atmosphere packaging of fresh fruits and vegetables: Modeling with tomatoes and oxygen absorber. *Journal of Food Science 68*(5), 1736–1742.

Charles, F., Marcano, J. S., & Gontard, N., (2005). Modeling of active modified atmosphere packaging of endives exposed to several postharvest temperatures. *Journal of Food Science, 70*(8), e443–e449.

Fante, C. A., Boas, A. C. V., Paiva, V. A., Pires, C. R. F., & Lima, L. C. O., (2014). Modified atmosphere efficiency in the quality maintenance of Eva apples. *Food Science and Technology, 34*(2). http://dx.doi.org/10.1590/fst.2014.0044.

Hegazy, R., (2013). *Post-Harvest Situation and Losses in India*. Technical Report. doi: 10.6084/m9.figshare.3206851.v1.

Helland, H. S., Gunnar, A. L., Bengtsson, B., Skaret, J., Lea, P., & Wold, A. B., (2016). Storage of fresh-cut Swede and turnip in modified atmosphere: Effects on vitamin C, sugars, glucosinolates and sensory attributes. *Postharvest Biology and Technology, 111*, 150–160.

Hyun, J. E., & Lee, S. Y., (2017). Effect of modified atmosphere packaging on preserving various types of fresh produce. *Journal of Food Safety*. doi: 10.1111/jfs.12376.

Kader, A. A., (2002). *Post-Harvest Technology of Horticultural Crops* (p. 535). Oakland: the University of California, Division of Agriculture and Natural Resources Publication 3311.

Kitsiou, S., & Sfakiotakis, E., (2003). Modified atmosphere packaging of 'Hayward' kiwifruit: Composition of the storage atmosphere and quality changes at 10°C and 20°C. *Acta Horticulture, 610,* 239–244. doi: 10.17660/Acta Hortic.2003.610.31.

Magazin, N., Keserović, Z., Cabilovski, R., Milic, B., Doric, M., & Manojlović, M., (2015). Modified atmosphere packaging of fully ripe strawberries. *Acta Horticulturae, 1071*(1071), 241–244. doi: 10.17660/Acta Hortic.2015.1071.28.

Manolopouloul, & Varzakas, T. H., (2013). Effect of modified atmosphere packaging (MAP) on the quality of 'ready-to-eat' shredded cabbage E. *International Journal of Agricultural and Food Research, 2*(3), 30–43. ISSN: 1929-0969.

Mozhdehi, F. J., Sedaghat, N., & Seyed, A. Y. A., (2017). Effect of modified atmosphere packaging (MAP) on the moisture and sensory property of saffron. *MOJ Food Process Technology, 5*(1), 212–214. doi: 10.15406/mojfpt.2017.05.00115.

Mudau, A. R., Soundy, P., Araya, H. T., & Mudau, F. N., (2018). Influence of modified atmosphere packaging on postharvest quality of baby spinach (*Spinacia oleracea* L.) leaves. *Horticulture Science, 53*(2), 224–230.

Page, N., González-Buesa, J., Ryser, E. T., Harte, J., & Almenar, E., (2016). Interactions between sanitizers and packaging gas compositions and their effects on the safety and quality of fresh-cut onions (*Allium cepa* L.) *International Journal of Food Microbiology, 218*, 105–113.

The Economic Survey, (2016–2017). *Agricultural and Processed Food Products Export Development Authority (APEDA).* Department of Commerce and Industry, Union Budget 2017–18, Press Information Bureau, Ministry of Statistics and Program Implementation, Press Releases, Media Reports, Ministry of Agriculture and Farmers Welfare, Crisil.

Watkins, C. B., & Nock, J., (2012). *Production Guide for Storage of Organic Fruits and Vegetables. New York State Department of Agriculture and Markets, Publication #10*, http://nysipm.cornell.edu/organic_guide/stored_fruit_veg.pdf (accessed on 19 December 2020).

Yang, H. C., Qing, W. J., Xin, S. B., & Fan, Y., (2012). Study on modified atmosphere packaging of fresh spinach at room temperature. *Packaging Engineering.*

CHAPTER 8

STORAGE OF FRUITS AND VEGETABLES: AN OVERVIEW

SOUMITRA BANERJEE,[1] SWARRNA HALDAR,[1]
B. N. SKANDA KUMAR,[1] and JAYEETA MITRA[2]

[1]*Centre for Incubation, Innovation, Research, and Consultancy (CIIRC), Jyothy Institute of Technology, Bangalore, Karnataka, India*

[2]*Agricultural and Food Engineering Department, Indian Institute of Technology Kharagpur, West Bengal, India,*
E-mail: jayeeta.mitra@agfe.iitkgp.ernet.ac.in

ABSTRACT

India stands second in fruits and vegetable production in the world. According to National Horticulture Database (National Horticulture Board, India), fruit production was 86.602 million metric tons, and vegetable production was 169.478 million metric tons in the year 2014–15. Vast climatic diversity allows India to grow diverse varieties of fruits and vegetables (F&V), but storage of these agricultural commodities is a challenge since these commodities are known for their highly perishable behavior. F&V grown in tropical climatic regions tend to deteriorate rapidly due to high temperature and humidity. Cultivated F&V are consumed/processed locally or exported to other states or countries. For all the purposes, it is required to store the raw agricultural produce for a targeted time without any spoilage. Storage plays an important role in enhancing the shelf life and preventing spoilage of F&V. The aim of this book chapter is to give the reader a concise knowledge about the storage of various F&V, different types of storage structures and conditions adopted for different F&V, types of storage structures and changes taking place during storage of F&V. The concept of Zero Energy Cooling Chamber (ZECC) is discussed herein brief as an economical storage solution for storage of F&V.

8.1 INTRODUCTION

Fruits and vegetables (F&V) are quite similar in many aspects of their composition, harvesting methods, storage, postharvest processing, etc. Scientifically fruits are those parts of the plant that contains the seeds of the plants for future reproduction, whereas vegetables come from the other plant part. Based on this distinction, eggplant, tomatoes, okra, etc., can be called fruits, whereas they are commonly referred to as vegetables. But in day to day life, whatever plant sources are consumed as main course meal are called vegetables and those which are eaten as a dessert or alone are called fruits. Vegetables are obtained from various plants parts like roots, leaves, stems, buds, etc., based on which they can be classified as shown below (Srivastava and Kumar, 1994; Potter and Hotchkiss, 2012):

- **Under-Earth Vegetables:** Vegetables, that grows under the earth, which may be classified as roots, stems, and bulbs:
  - **Roots:** Carrot, radish, beetroot, turnip, arracacha.
  - **Stems:** Taro.
  - **Tubers:** Potato, sweet potato, tapioca.
  - **Bulbs:** Onion, garlic, leek.
- **Leafy Vegetables:** These are the leafy part of the plant that grows above the ground, which are classified as leaves, petioles, flower buds and shoots:
  - **Leaves:** Lettuce, spinach.
  - **Petioles:** Rhubarb.
  - **Buds:** Cauliflowers.
  - **Shoots:** Asparagus.
- **Cole Crops:**
  - Cabbage, cauliflower, khol-khol, Brussel sprouts, Chinese cabbage, broccoli.
- **Fruit-Vegetables:** These are those vegetables, which are technically considered as fruits, but eaten as vegetables:
  - **Legumes:** Peas, beans.
  - **Cereals:** Corn.
  - **Vine:** Cucumber.
  - **Berry:** Tomato, eggplant.
  - **Tree Fruit:** Breadfruit.

Fruits, on the other hand, are the matured plant ovaries with seeds, known for their juicy acidic-sweet taste. Fruits are classified based on structure, chemical composition, climatic requirements, etc. Some classifications of fruits are given below:

- **Berries:** Strawberries, cranberries, grapes.
- **Melons:** Watermelon, honeydew.
- **Drupes:** Apricot, cherries, peaches, plums.
- **Pomes:** Apples, quince, pears.
- **Citrus Fruits:** Oranges, lemons.
- **Tropical/Sub-Tropical Fruits:** Banana, dates, figs, pineapples.

Composition wise F&V are quite similar since both of them contain high moisture content and low in fat and proteins. Vegetables and fruits are a good source of digestible and indigestible carbohydrates, which are essential for the human body. Beside these macronutrients, fruits, and vegetables contain other essential micronutrients, which are also essential for human nutrition, i.e., minerals, vitamins, and other important phytochemicals. F&V being high moisture commodities and enriched in nutrients, are highly susceptible to microbial, chemical, or other spoilage. It is desirable to take necessary steps and precautions to extend the shelf life of the harvested products by proper storage (Srivastava and Kumar, 1994; Potter and Hotchkiss, 2012; APEDA, 2018).

As reported by World Farmers Organization (WFO, 2014), agricultural market information service (AMI) of Bonn reported that worldwide 1.74 billion tons of F&V were cultivated in the year 2013, which is 9.2% more than that of the previous year. Individually, the production of F&V were 950 and 790 million tons, respectively (WFO, 2014). India is a country of diverse climatic zones, allowing farmers to cultivate almost every type of F&V. India ranks second in global fruits and vegetable production and China ranks first. As reported by APEDA (2018), National Horticulture Board (India) reported that during 2014–15, fruits, and vegetables produced in India were 86.602 and 169.478 million tons, respectively. For fruits and vegetable cultivation, India has cultivation areas of 6.110 and 9.542 million hectares, respectively. This huge production of F&V allows lucrative export opportunities around the globe like Malaysia, Saudi Arabia, United Arab Emirates (UAE), Bangladesh, Sri Lanka, Nepal, UK, Malaysia, Pakistan, and Qatar, etc. A growing acceptance trend in Indian Horticulture products has been observed because of developments in storage facilities, cold chain management, quality assurance, etc., (Srivastava and Kumar, 1994; Shakuntala and Shadaksharaswamy, 2001; Potter and Hotchkiss, 2012; APEDA, 2018).

8.1.1 POST-HARVEST LOSSES

Plant-based food sources like F&V contain high moisture content and continue respiration process even after harvest. This caused moisture losses and usage of nutrients present in harvested F&V, which causes nutrient and moisture losses along with other undesirable effects causing the product to decay with time. Besides internal factors, external factors are also responsible for decay and food losses, i.e., temperature, relative humidity (RH), etc. These internal and external factors cause food losses and wastage. These postharvest losses may happen at any point between the harvesting of agriculture produce, transport, storage, and its final distribution to the consumer. Types of postharvest damages may vary from physical damages, physiological decay, moisture loss, microbial/pest/ rodent contamination, etc. (Burden and Wills, 1989; FFTC, 2018).

8.1.2 FACTORS RESPONSIBLE FOR SPOILAGE

For designing better storage conditions of F&V, it is essential to know about various factors responsible for F&V spoilage. By considering effective measures to prevent the spoilage factors, it is possible to design safer storage structures for F&V. All the major factors responsible for the spoilage of F&V may be categorized into the following five categories as shown in Table 8.1 (Srivastava and Kumar, 1994; Chakraverty, 2005):

TABLE 8.1 Major Factors Responsible for Food Spoilage

SL. No.	Type of Factors	Factors
1.	Biological	Microorganisms, insects, rodents
2.	Physiological	Enzymes, respiration
3.	Physical	Temperature, relative humidity
4.	Chemical	Moisture, oxygen
5.	Other factors	Mechanical damages

All the factors mentioned in Table 8.1 are explained in the below paragraphs:

1. **Biological Factors:**

   i. **Microbial Spoilage:** After harvesting of F&V, primary microorganisms that play significant roles in the spoilage are bacteria, yeast, and molds. These microorganisms are present almost everywhere, including air, soil, water, etc. Bacteria are unicellular microorganisms of 1 μm dimension. Bacteria can be of different shapes like spherical, cylindrical, spiral, etc. Their method of reproduction is by "cell division." The important genus of food spoilage bacteria is *Acetobacter, Gluconobacter, Lactobacillus, Leuconostoc, Pediococus, Clostridium, Propionibacterium, Proteus*, etc. The optimal growing temperature for bacterial growth is at 37°C. For destroying bacterial spore forms high temperature of 100°C or above is required. Bacteria are sensitive to lower pH, hence sterilization of acidic fruits are possible at 100°C whereas, for sterilization of higher pH alkaline vegetables, higher temperature above 100°C is required (Srivastava and Kumar, 1994; Jay, 2000).

Similar to bacteria, yeast are unicellular microorganisms but are larger in dimension, i.e., 10 μm, unlike bacteria. Yeast is of spherical or ellipsoidal shaped organisms, which multiply by the method of "budding." They grow well in the presence of abundant moisture and low sugar, for suitable fermentation to produce carbon dioxide and alcohol. Boiling destroys yeast cells and spores. Higher sugar concentration acts hindrance to yeast growth. Yeasts responsible for spoilage of fruits are *Saccharomyces, Candida, Brettanomyces, Pichia, Torulopsis*, and *Rhodotorula*, etc. Unlike bacteria and yeast, molds are multicellular and filamentous fungi and are larger in structure than yeast. They grow well in close, damp, dark places and require much lesser moisture than yeast and bacteria. Sugar-containing fruit products like jam and jellies are spoiled because of molds. Molds grow well in acid media as well but are not found in alkaline media. Boiling destroys both mold and its spores. Molds related to food spoilage are *Penicillium, Aspergillus, Mucor, Byssochlamys fulva*, etc. Few molds are responsible for toxic products in foods, i.e., mycotoxins. The conventional method of prevention of microbial spoilage of F&V are listed below (Srivastava and Kumar, 1994; Jay, 2000; Potter and Hotchkiss, 2012):

a. Altering storage environment to prevent/reduce microbial activity.
b. Thermal treatment, i.e., sterilization.
c. Acidification.
d. Chemical methods, i.e., lactic, citric, tartaric acids.
e. Antimicrobial agents, i.e., sulfur dioxide, benzoic acid, sorbic acid, etc.
f. Antioxidants like ascorbic acid, butylated hydroxy-anisole, erythrobic acid, etc.

ii. **Insect and Rodent Spoilages:** Insects cause massive destruction of F&V depending on 5–50% of total produce depending according to the precaution taken during production in field and storage. Insects cause spoilage in F&V in two ways, i.e., firstly by direct consumption of the agricultural produce and secondly by causing damage by bruises and cuts on the food surface allowing initiation of microbial activities. Parasites are also responsible for spoilage of F&V, which get spread in the agricultural produce by infected water and poor hygienic practices adopted during food production or storage (Srivastava and Kumar, 1994; Chakraverty, 2005; ICAR, 2012).

Rodents like rats also cause damage to F&V in field and storage structures if their population is not controlled effectively. Rats are known to produce 3–8 offsprings in their life span of 3 years. They cause spoilage in three ways, i.e., direct consumption of agriculture produce, damaging the fruits and vegetable surface, hence encouraging microbial spoilage and by contaminating the stored foods with their urine and droppings which causes development of various deadly diseases like typhus, plague, typhoid, etc., (Srivastava and Kumar, 1994; Chakraverty, 2005; ICAR, 2012).

2. **Physiological Factors:**

i. **Enzymatic Spoilage:** Enzymes are responsible to initiate many reactions in food systems like ripening and spoilage. This enzyme may come from the fruits or vegetables directly or could get produced by microbial contamination. Optimum enzyme initiated reaction temperature is 37°C, and at 80°C all enzymes gets deactivated. Enzymatic actions can be prevented safely by controlling storage temperature and air moisture content. Browning of F&V are caused by enzymes like phenolase, peroxidase, polyphenol oxidase enzymes (Potter and Hotchkiss, 2012; Srivastava and Kumar, 1994).

ii. **Respiration:** Agricultural commodities being live foods, respires even after harvesting. During complete aerobic respiration, carbohydrate (hexose group) is broken down to yield carbon dioxide and water with release of energy to increase the storage room temperature. Respiration causes loss of mass, gain in moisture, the rise in grain temperature, etc. All these factors promote spoilage of the agriculture produce. F&V can be categorized based on respiration, as shown in Figure 8.1 (Gast, 2001). Respiration rate of freshly harvested F&V can be reduced by either reducing the temperature, oxygen content or by increasing the carbon dioxide content inside the storage room (Potter and Hotchkiss, 2012).

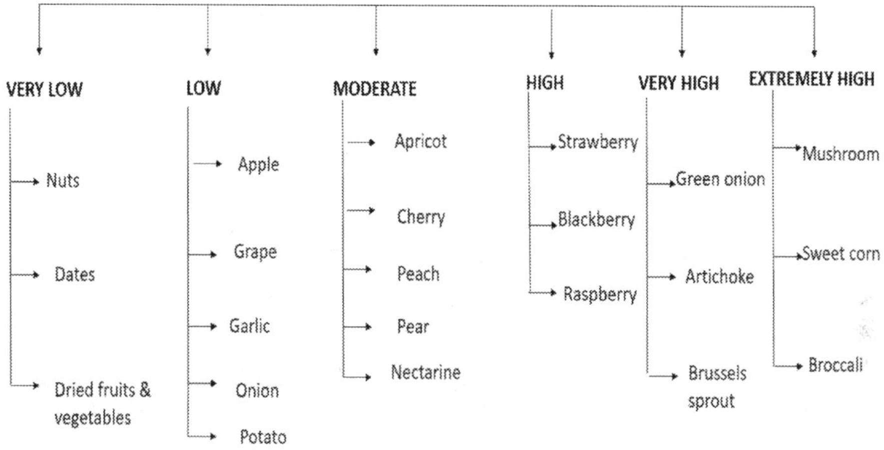

VERY LOW
- Nuts
- Dates
- Dried fruits & vegetables

LOW
- Apple
- Grape
- Garlic
- Onion
- Potato

MODERATE
- Apricot
- Cherry
- Peach
- Pear
- Nectarine

HIGH
- Strawberry
- Blackberry
- Raspberry

VERY HIGH
- Green onion
- Artichoke
- Brussels sprout

EXTREMELY HIGH
- Mushroom
- Sweet corn
- Broccali

FIGURE 8.1 Classification of fruits and vegetables according to their respiration rates.
Source: Adapted and modified from Gast (2001); and Silva (2010).

3. Physical Factors:

i. **Temperature:** For F&V, storage temperature has an important role. Ripening and spoilage of agricultural produce get accelerated when the temperature is not controlled. Temperature assisted chemical reactions doubles its rate for every 10°C rise in temperature. Excessive higher temperature storage causes protein denaturation, vitamin losses, moisture loss, color changes, texture deformation, etc. Reduced temperature increases the storage shelf life of stored F&V. Although it has been seen that for some F&V stored at the reduced temperature of 4°C, gets deteriorates. For example, banana, lemon, squash can be stored for a longer duration, when stored at 10°C. Very lower temperature, i.e., below freezing temperature (–18°C) prevents microbial growth and may also cause reduction of microbial load, but may cause chilling injuries to fruits. It is necessary to evaluate the characteristics of the food, before deciding its storage temperature (Srivastava and Kumar, 1994; Gast, 2001).

ii. **Relative Humidity (RH):** It is the ratio of partial pressure of water vapor in the air to the partial pressure of water vapor in saturated air under the same temperature. It is the measure of moisture present in the air expressed in percent, where 100% RH means air fully saturated with moisture and 0% RH means absolute dry air. Moisture gets evaporated from freshly harvested F&V if storage air RH is too low. Higher RH, on the other hand, reduces moisture loss from the harvested products and keeps appearance, nutritional quality stable. It is essential to monitor and control temperature as well as humidity inside a storage chamber. Moisture transmission rate is not only decided by RH alone but other factors like temperature, air velocity, and atmospheric pressure (Srivastava and Kumar, 1994; Gast, 2001; Potter and Hotchkiss, 2012).

4. **Chemical and Other Factors:**
   i. **Chemical Factors:** Oxygen plays a significant role in food spoilage and is responsible for undesirable changes in food color, flavors, etc. Oxygen promotes the growth of various microorganisms and controlling the oxygen level by deaeration, vacuums packaging, changing the gas environment by introducing inert gas like nitrogen or carbon dioxide, etc.
   ii. **Mechanical Damage:** During careless or inexperienced methods of harvesting or rough handling causes injuries of F&V. Even slight bruises or slight surface damages help microbial contamination or other deterioration to start. It is recommended to remove the damaged or injured products during grading or postharvest treatment (Srivastava and Kumar, 1994).
   iii. **Other Factors:** The other factors responsible for spoilage of F&V are light and storage duration. High-intensity direct light is responsible for the destruction of Vitamins B_2, A, and C and have effects on other food parameters like the color. Light sensitive foods may be stored in improvised packaging materials.

Last factor responsible for food deterioration is storage duration. Harvested food stored for the longer duration of time, would have more chances to spoilage. It is better for the harvested agricultural commodity to be either consumed or processed to prevent any deterioration (Srivastava and Kumar, 1994; Potter and Hotchkiss, 2012).

8.1.3 FOOD STORAGE, FOOD SECURITY, AND pH LOSSES PREVENTION

Human population growth, with the increase in food demands, is rising annually. Agriculture has been experiencing development in technological and scientific advancement, for the improvement of productivity, higher yield, and better food crop production. Still, human agricultural development has not reached the point, to provide sufficient nutrients to the overall global population. There is a reduction of global hunger level, but still, many people are not in the position to afford sufficient nutrients from their diet, leading them to suffer from nutritional deficiencies. Modern days farmers are facing new challenges for agricultural land, water, energy, climate changes, etc., which also sometimes causes negative effects on the agricultural productivities. Growing demand for food and upcoming challenges makes food sustainability a modern-day challenge, with rapid growing human population. This food sustainability challenges may be solved with better scientific and technological intervention and their adaptation in modern agricultural systems, to prevent the challenges, with multifaceted and linked global strategy for ensuring food safety and sustainability (Godfray et al., 2010). World Food Summit (1996) under United Nations, defined food security exists, "when all people, at all times, have physical, social, and economic access to sufficient, safe, and nutritious food to meet dietary needs for a productive and healthy life." Food security is a major global challenge due to the factors like global climatic changes, population growth, water problems, energy shortages, etc. Food security may be ensured by adaptation of proper strategic planning and formulating policies, for better agricultural practices, agricultural land utilization, postharvest processing, pricing, and distribution of foods (IFPRI, 2018). Different researchers and organizations at different times have defined food security (Maxwell and Smith, 1992). Food storage plays important roles in ensuring longer storage of foods without spoilage and thereby reducing Post Harvest (PH) losses. Postharvest losses can be reduced by careful selection of storage condition because optimal storage condition would improve the shelf life of the storage agriculture produce. By extending the shelf life, the F&V could be stored for a longer duration without much chances of spoilage.

8.2 STORAGE CONDITIONS FOR DIFFERENT FRUITS AND VEGETABLES (F&V)

Freshly harvested F&V are sensitive and are susceptible to spoilage. Factors affecting their conditions and shelf life, i.e., temperature, environment, etc., has already been discussed. It is well understood that F&V need to be stored in an optimal storage condition to increase the shelf life and quality of the harvested F&V. Now let us understand the specific importance of storages for F&V, as given in Figure 8.2 (ICAR, 2012).

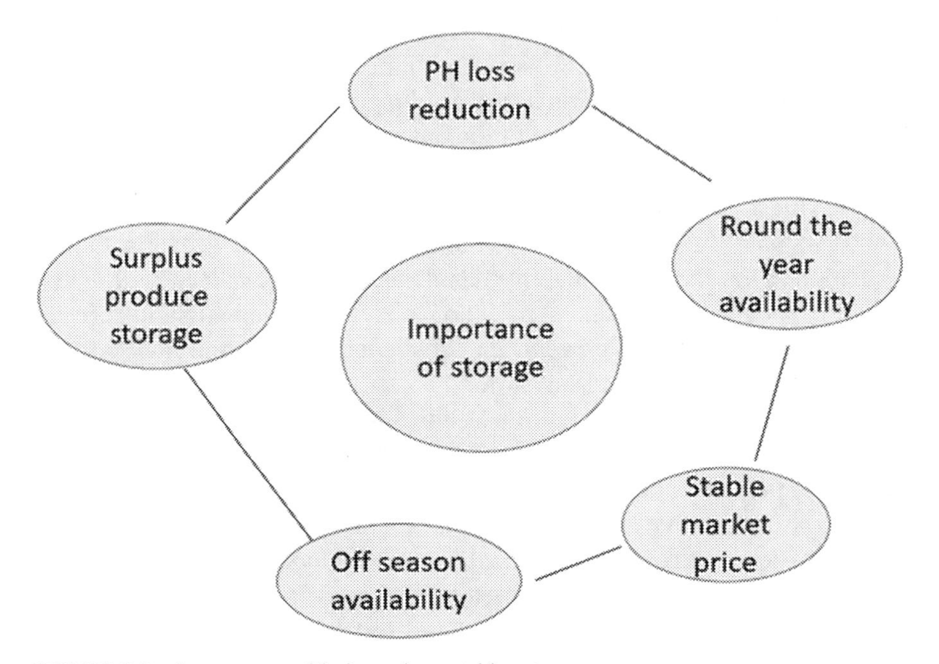

FIGURE 8.2 Importance of fruits and vegetables storages.
Source: Adapted and modified from Chakraverty (2005); and ICAR (2012).

From Figure 8.2, we can understand the important purposes of storage of F&V. Storage conditions of different F&V are shown in Tables 8.2 and 8.3. For storage of F&V, conventionally temperature and RH are controlled as can be seen in Tables 8.2 and 8.3.

8.3 TYPES OF STORAGES FOR FRUITS AND VEGETABLES (F&V)

Till this part of the discussion, we can conclude that F&V are quite sensitive to spoilage, and as they are harvested, they should be handled and stored in an optimal storage condition. Different foods require different storage conditions for different purposes. Here we would be discussing the five popular types of storages which are conventionally adopted for the storage of F&V, as mentioned below (ICAR, 2012). All these different types of storages mentioned below would be discussed in brief:

- Low-temperature storages;
- Controlled atmospheric storage;
- Hypobaric storage;
- ZECC structures.

TABLE 8.2 Optimal Storage Condition for Fruits

Name of Fruits	Optimal Conditions	Shelf Life	References
Apples	Temp.: 2–3°C	16–17 weeks	ICAR (2012)
	Temp.: –1 to –4°C RH (%): 90–95	1–12 month	Gast (2001); FAO (2004); El-Ramady et al. (2015)
Bananas	Temp.: 12–13°C	3–4 weeks	ICAR (2012)
Banana-Platinum	Temperature: 13–15°C RH (%): 90–95	7–28 days	FAO (2004); El-Ramady et al. (2015)
Grapes	Temp.: 0–2°C	5–7 weeks	ICAR (2012)
Grapes (American)	Temp.: –0.3°C RH (%): 85	2–8 weeks	Gast (2001)
Guavas	Temp.: 8–10°C	2–5 weeks	ICAR (2012)
Jackfruits	Temp.: 11–13°C	7–8 weeks	ICAR (2012)
Limes	Temp.: 11–13°C	8 weeks	ICAR (2012)
Mangoes	Temp.: 11–13°C	4 weeks	ICAR (2012)
Oranges	Temp.: 11–13°C	8 weeks	ICAR (2012)
Sapota	Temp.: 19–21°C	3 weeks	ICAR (2012)
Tomatoes (mature)	Temp.: 16–20°C	3 weeks	ICAR (2012)
Tomatoes (mature-green)	Temp.: 12.7–21°C RH(%): 90–95%	1–3 weeks	Gast (2001)
Tomatoes (firm-ripe)	Temp.: 12.7–21°C RH(%): 90–95%	4–7 days	Gast (2001)
Pineapples	8–10°C	6 weeks	ICAR (2012)
Blackberries	Temp.: –0.28°C RH (%): 90–95	2–3 days	Gast (2001)
Elderberries	Temp.: –0.28°C RH (%): 90–95	1–2 weeks	Gast (2001)
Gooseberries	Temp.: –0.28°C RH (%): 90–95	3–4 weeks	Gast (2001)
Raspberries	Temp.: –0.28°C RH (%): 90–95	2–3 days	Gast (2001)
Strawberries	Temp.: –0.28°C RH (%): 90–95	3–7 days	Gast (2001)

TABLE 8.2 *(Continued)*

Name of Fruits	Optimal Conditions	Shelf Life	References
Nectarines	Temp.: –0.28°C RH (%): 90–95	2–4 weeks	Gast (2001)
Peaches	Temp.: –0.28°C RH (%): 90–95	2–4 weeks	Gast (2001)
Pears	Temp.: –0.28°C RH (%): 90–95	2–7 months	Gast (2001)
Plums and prunes	Temp.: –0.28°C RH (%): 90–95	2–5 weeks	Gast (2001)
Quinces	Temp.: –0.28°C RH (%): 90–95	2–3 months	Gast (2001)

TABLE 8.3 Optimal Storage Condition for Vegetables

Name of Commodity	Optimal Conditions	Shelf Life	References
Beans	Temp.: 0–2°C	3 weeks	ICAR (2012)
Beans (dry)	Temp.: 4.5–10°C RH (%): 40–50	6–10 months	Gast (2001)
Beans green or snap	Temp.: 4.5–10°C RH (%): 95	7–10 days	Gast (2001)
Beans (lima)	Temp.: 2.78–5°C RH (%): 95	5–7 days	Gast (2001)
Beans (sprouts)	Temp.: 0°C RH (%): 95–100	7–9 days	Gast (2001)
Beetroot	0–2°C	7 weeks	ICAR (2012)
Beets (bunched)	Temp.: 0°C RH (%): 98–100	10–14 days	Gast (2001)
Beets (topped)	Temp.: 0°C RH (%): 98–100	4–6 months	Gast (2001)
Brinjal	Temp.: 8–10°C	4 weeks	ICAR (2012)
	Temp.: 7.8–12°C RH (%): 90–95	1 week	Gast (2001)
Cabbage	0–2°C	12 weeks	ICAR (2012)
Cabbage (early)	Temp.: 0°C RH (%): 98–100	3–6 weeks	Gast (2001)
Cabbage (late)	Temp.: 0°C RH (%): 98–100	5–6 months	Gast (2001)
Cabbage (Chinese)	Temp.: 0°C RH (%): 95–100	2–3 months	Gast (2001)

TABLE 8.3 *(Continued)*

Name of Commodity	Optimal Conditions	Shelf Life	References
Cauliflower	Temp.: 0–2°C	7 weeks	ICAR (2012)
	Temp.: 0°C RH (%): 95–98	3–4 weeks	Gast (2001)
Capsicum	Temp.: 11–13°C	3 weeks	ICAR (2012)
Cucumber	Temp.: 7–8°C	2 weeks	ICAR (2012)
	Temp.: 10–12.8°C RH (%): 95	10–14 days	Gast (2001)
Ginger	Temp.: 2–3°C	14 weeks	ICAR (2012)
Indian gooseberry	Temp.: 2–3°C	8 weeks	ICAR (2012)
Knol-khol	Temp.: 0–2°C	12 weeks	ICAR (2012)
Okra	Temp.: 8–10°C	2 weeks	ICAR (2012)
	Temp.: 7.2–10°C RH (%): 90–95	7–10 days	Gast (2001)
Onions	Temp.: 0–2°C	24 weeks	ICAR (2012)
Onions (green)	Temp.: 0°C RH (%): 95–100	3–4 weeks	Gast (2001)
Onion (dry)	Temp.: 0°C RH (%): 65–70	1–8 months	Gast (2001)
Onion (Sets)	Temp.: 0°C RH (%): 65–70	6–8 months	Gast (2001)
Potatoes	Temp.: 2–3°C	34 weeks	ICAR (2012)
Potatoes (early crop)	Temp.: 4.5°C RH (%): 90–95	4–5 months	Gast (2001)
Potatoes (late crop)	Temp.: 3.4–4.5°C RH (%): 90–95	5–10 months	Gast (2001)
Snake gourd	Temp.: 18–21°C	2 weeks	ICAR (2012)
Tapioca	Temp.: 0–2°C	23 weeks	ICAR (2012)

8.3.1 REFRIGERATION AND COLD STORAGES

During storage of F&V, various spoilage factors start acting against the stored foods, which initiates the deterioration process, i.e., microbial and enzymatic spoilages. By reducing the temperature, it is possible to arrest the deterioration process. Lower the temperature lower would be the deterioration rate. Temperature management is an effective method for stabilizing the agricultural products and keeps them for a longer duration of time, without any spoilage. Low-temperature preservation methods are one of the oldest methods, but after the invention of mechanical ammonia refrigeration in 1875, commercial preservation of foods in refrigerated chambers came to the picture. Mechanical refrigeration played a major role in the world of agricultural trades from farm to consumers. Refrigeration and cold storage are approved methods of food storage to make it available around the year, stabilize market price and prevent food wastages. Cold storage usually denotes storage structures where temperature ranges from 16 to –2°C. Domestic refrigerators and commercial refrigerators operate at 4–7°C, whereas frozen storage refrigerators operate at the lower temperature of –18°C or below. The shelf life of the preserved foods under low temperature would depend on the quality of the foods stored. Under cold storage, perishable foods can be stored for few days to weeks whereas, under the frozen condition, same foods can be stored for months and even years with proper packaging and other conditions associated (Potter and Hotchkiss, 2012).

Lower temperature storage is considered as the novel method of food preservation since there are least deteriorative effects on taste, texture, nutritional composition, and other food quality aspects when food is stored for a recommended duration. It is a well-established fact that lowering the temperature reduces the microbial and other undesirable chemical reaction kinetics. Although it has been reviewed by several researchers that microbial growth rate reduces significantly at low temperatures, studies have claimed that few pathogenic microorganisms have the capability to grow even at the temperature of 3.3°C. Particularly below –9.5°C, microbial activity ceases, and reduction in microbial population could be experienced (Potter and Hotchkiss, 2012).

8.3.2 CONTROLLED ATMOSPHERIC STORAGES

Inside storage rooms, the F&V utilize the atmospheric oxygen present and give off carbon dioxide as their respiration process. Atmospheric air contains a high amount of oxygen (21% approx) which is responsible for oxidation reaction, microbial and other organisms' growth. Altering the gaseous composition of the surrounding atmosphere inside the storage space affect the stored F&V. The reduced concentration of oxygen along with the increased composition of carbon dioxide reduces the respiration rates of agricultural products along with inhibitory effects on microbial and insect growth.

Controlled atmospheric storages have controlled composition of gases like O_2 and CO_2 inside the storage chamber. The gaseous composition is maintained by monitoring the inner storage gas composition. It has already been discussed about the respiration phenomena of F&V, due to which the composition of gas inside the storage changes dynamically. For keeping the desired gas composition constant, pre-determined levels of fresh gas flushing or chemical absorbent/adsorbents are used to remove excess amount of undesirable gases from the storage structure (ICAR, 2012). Plant cells are responsible for the production of ethylene, which is responsible for ripening and other negative effects like breakdown of chlorophyll, softening of kiwi fruits, etc. Chemical absorbents can be used to remove or absorb out the ethylene. Lawton (1991) reported that ethylene can be removed by maintaining proper air ventilation, potassium permanganate, heat-treated platinum catalyst and UV radiation (184–254 nm). Fruits have a lower shelf life and their storage in controlled atmospheric storage increases the shelf life. For example, McIntosh apples can be stored for 1 month at 3°C temperatures, 87% RH, 1% oxygen content, 3% carbon dioxide content. In some other cases, the concept of controlled atmospheric storage is used to incorporate antimicrobial fumigants to control microbial activities (Potter and Hotchkiss, 2012).

8.3.3 HYPOBARIC STORAGE

This storage can be called a modified form of CA storage, where low-temperature storage structures are maintained under reduced pressure with high humidity. Low pressure reduces the storage air content and thus reduces the net oxygen content. Higher RH is maintained inside the storage structure to ensure prevention of moisture dehydration from the stored F&V. This kind of storage reduces the microbial and enzymatic spoilage rates and is considered energy efficient, as this storage structure requires lesser intensity refrigeration (Potter and Hotchkiss, 2012). Chem et al. (2013) studied the effects of hypobaric storage of Chinese bayberry fruits and reported a reduction of fruit decay and total acid losses. Inhibitory respiratory effects were observed along with not much change in total phenolic content and antioxidant activity. The study inferred hypobaric storages were recognized as effective postharvest storage technology for extending shelf life. In another study on comparative effects of green asparagus stored in hypobaric storage condition, it was found that storage life gets extended to 50 days, whereas under refrigeration and room condition, storage life was found for 25 and 6 days respectively, without compromising the product quality factors like chlorophyll, vitamin C, titratable acidity and soluble solids, etc.

8.3.4 *ZERO ENERGY COOLING CHAMBERS*

Zero energy cool chamber (ZECC) or evaporative cooling chamber is simple and low-cost storage structures, build using locally available raw materials like bricks, sand, bamboo, jute cloth, riverbed sand, dry grass, etc. Inside storage, temperature ranges between 10 and 15°C lower with RH ranging from 30–40% higher than the atmosphere. Further studies confirmed the safer storage of F&V for more than 3–5 days compared to room temperature storage. Economical evaporative storage structures require no external energy sources and can be constructed with locally available materials and unskilled labor force (Roy and Pal, 1989; Dadhich et al., 2008). Pusa Zero Energy Cold Chambers (ZECC) was developed and technology was transferred to the farmers by IARI, under the sponsorship of National Horticulture Board IARI, under the sponsorship of National Horticulture Board, UP Diversified Agriculture Support Project, Ministry of Food Processing Industry (MOFPI). ZECC had specific advantages like easy to construct, with locally available materials. There is no external power requirement and allows the farmers to store the agriculture commodities fresh for few days after harvest (IARI, 2018).

8.4 CHANGES TAKING PLACE DURING STORAGE OF FRUITS AND VEGETABLES (F&V)

During the harvesting time, many physiological changes occur in plant tissues. The supply of water, hormones, vitamins, mineral, and organic molecules is hindered after the F&V are detached from the parent plant. There is little or no new photosynthesis, but the plant tissues transform most of the constituents which are already present (Ludford, 1987). Immediately after harvest, sensorial, nutritional, and organoleptic quality of fresh produce starts deteriorating as a result of altered plant metabolism and microbial growth. The quality deterioration is the result of transpiration, senescence, ripening-associated processes, wound-initiated reactions, development of postharvest disorders, microbial proliferation leading to postharvest quality loss. Rates of compositional changes (associated with color, texture, flavor, and nutritive value), mechanical injuries, water stress, sprouting, and rooting, physiological disorders, and pathological breakdown also causes quality and postharvest losses. The rate of biological deterioration depends on several environmental (external) factors, including temperature, RH, air velocity, and atmospheric composition (concentrations of oxygen, carbon dioxide, and ethylene), and sanitation procedures. Postharvest changes occurring in F&V during storage are discussed in subsections.

8.4.1 APPEARANCE

Ripening of F&V changes the skin color after postharvest. The green color of unripe fruits like apples, banana, papaya, tomato, etc. becomes lighter due to the breakdown of chlorophyll and formation of other pigments like yellow, orange, red, etc., (Tucker, 1993). Other appearance changes like loss of freshness viz., wilting of leaves (leafy vegetables), loss of glossiness or surface wrinkling also gradually sets in after postharvest.

8.4.2 TEXTURE

Cell wall degradation causes crisp firm tissues of fresh produce to become soft because postharvest (Martínez-Romero et al., 2007), although tissue softening is desirable for optimal quality of fruit but over-ripening is a sign of senescence or internal decay. Also in some stem crops like asparagus, tough fibers are developed during storage.

8.4.3 FLAVOR AND AROMA

Flavor is a complex blend of two components: aroma components and taste. Important taste components in freshly produced F&V are bitterness, sweetness, astringency, and acidity. Organic acids decrease during storage. Sugar level increases during ripening in some fruits like mangoes, bananas, pears, apples, etc., due to the conversion of starch into sugar. This eventually increases the sweetness of the fruits. However, if the acid to sugar ratio falls considerably, then the F&V becomes bland. It is important to maintain sugar/acid ratio in some fruits like grapes, limes, etc., to maintain proper flavor balance. Formation of many bitter components also occurs during storage components (Acked, 2002).

Aroma plays an important positive factor in some aromatic F&V like melons, mangoes, etc. Aroma profile of climacteric fruits changes dramatically after postharvest. The volatile components gradually change from unripe state to ripe state, over-ripe state and senescing state (Morton and

Macleod, 1990). Many unpleasant aromas are formed during improper storage due to bacterial and fungal contamination.

8.4.4 CHANGES DURING RESPIRATION

Every living tissue requires energy to maintain life. The energy is produced by respiration where carbohydrates are metabolized to produce carbon dioxide, water, and energy (ATP). Respiration occurs in cell mitochondria where it channels the released energy for maintenance of synthetic reactions, which take place after harvest. Respiration involves three enzymatic reactions viz.: (i) glycolysis (where glucose is converted to pyruvate); (ii) tricarboxylic acid cycle (where pyruvate is converted to carbon dioxide); and (iii) oxidative phosphorylation (where adenosine triphosphate, i.e., ATP is produced from reduced nicotinamide adenine dinucleotide and reduced flavin adenine dinucleotide). ATP is the main energy molecule and is needed for different cell functions. If respiration is reduced, then ATP production will also be reduced. When less energy is available during the ripening process, the quality changes (Dell et al., 2010).

Higher rate of respiration of fresh produce is associated with lower shelf life. For example, peas and beans have higher respiration rates and short shelf lives, whereas potatoes and onions have lower respiration rate and higher shelf lives. Some F&V have good source of energy reserves, e.g., potatoes, onion, carrots, that helps them to maintain respiration and energy production. However, some products like green leafy vegetables, broccoli, lettuce, endive, etc., energy reserve is less, and they are susceptible to excessive respiration and wilting.

8.4.5 CHANGES DURING TRANSPIRATION

About 80–90% of the mass weight of fresh F&V is water. Water content varies with the type of product. Water vapor and gases flow through a continuous pathway of intercellular spaces within the cells of F&V (Ben-Yehoshua, 1987). It is assumed that the water vapor and gases are in a saturated condition inside the cells (Sastry et al., 1978). However, internal bound water and solutes reduces the equilibrium moisture content from 100% and it becomes equivalent to 97% saturation vapor pressure of pure water (Watada et al., 1984). During transpiration, mass transfer takes place in the form of water vapor from the surface of the fruit and vegetables to the surrounding atmosphere. As a consequence of moisture loss, various phenomena like shrinkage, wilting, loss of firmness, and crispiness of F&V occurs. This eventually leads to undesirable changes in appearance, texture, and flavor. After losing 3–10% of mass, the freshness of fruits and vegetable diminishes (Burton, 1982). Transpiration is a major cause of postharvest losses in green leafy vegetables like spinach, lettuce, cabbage, chard, and green onion (Kader, 1983). Citrus fruits are also damaged due to transpiration (Ben-Yehoshua et al., 1985).

8.4.6 EFFECTS OF HORMONAL CHANGES

8.4.6.1 ETHYLENE

Ripening and senescence are triggered by ethylene production, which is a plant hormone (Reid, 1992). Ethylene is produced by all plant cells, and the production increases under stress conditions viz. pathogenic attack, physical damage, severe water loss, etc. Ethylene is produced in high amount during the start of the ripening process in climacteric fruits, and it controls the physical and chemical changes which are followed afterwards. Maturation and senescence are accelerated when exposed to exogenous ethylene. For example rate of chlorophyll loss in green vegetable increases; asparagus forms thick fibers; premature ripening in fruits and leaf abscission in cabbage and cauliflowers.

8.4.6.2 ABSCISIC ACID (ABA)

The pattern of change in abscisic acid (ABA) is not the same in all fruits. In some fruits like pear and avocado, free ABA level does not change but increases during ripening, indicating that they are ripening promoters (Rhodes, 1980). While in kiwi fruit, the content of ABA decreases during ripening (Tsay, 1984). In addition, in tomatoes, the hormone increases during the growth phase until the tomatoes reach "mature green" stage, and then it decreases gradually prior to ripening. An increase in free ABA level indicates an increase in net synthesis.

8.4.6.3 AUXINS

Auxins are also known as indoleacetic acid (IAA). The induction of ACC synthase by auxin influences ethylene production. It is thought that the endogenous auxin concentration is maximum after pollination and in the early stages of fruit development and minimum during maturation (Ludford, 1987). Some researchers suggested that auxin level and oxygen and ethylene interactions decreases during ripening and also changes in hydroperoxides takes place (Frenkel and Eskin, 1977). No correlation between No correlation between endogenous auxin content and fruit ripening has been found (McGlasson, 1978).

8.4.6.4 GIBBERELLINS (GAs)

Like auxins, GAs concentrations are high during the early stages of fruit development, and they may reduce senescence. GAs play an important role in color change. During the ripening process, chloroplasts are converted to chromoplasts. During winter, the rinds of Valencia oranges attain maximum orange color. GAs help them to return to their original green color during the spring and summer season (Thompson, 1967). In another study, it has been found that during tomato ripening, GA3 delays lycopene appearance (Dostol and Leopold, 1967).

8.4.6.5 CYTOKININS

Senescence in leaves fruit ripening is delayed by cytokinin treatment. As the ripening proceeds, the cytokinin level reduces. Active cell division after anthesis increases their level for two weeks in cherry tomatoes (Abdel-Rahman et al., 1975). Lower foliage to fruit ratio may induce increased cytokinin levels, thereby lowering sink competition in tomatoes (Varga and Bruinsma, 1974). Thus GAs, IAA, and cytokinins are involved in delay of ripening, while ABA levels may or may not increase during ripening.

8.4.7 SENESCENCE

Plant tissues grow old naturally and this natural process is called senescence. The factors which accelerate the respiration rate also speed up senescence, like increase in ethylene and other factors discussed above. During this process, all metabolic reactions stop and the plant tissues gradually dies. Senescence affects the physical and chemical integrity of the cell wall (Jimenez et al., 1997). Besides turgor pressure, cell wall integrity also affects the texture of many processed products (Femenia et al., 1998). During senescence, intercellular bonding between cells gradually weakens and breakdowns in some F&V (e.g., tomatoes and apples), the condition are called mealiness which results in loss in texture quality (Van der Valk and Donkers, 1994). While in potatoes, senescence is accompanied by the increase in sugar content from the stored starch. This is highly undesirable in potato processing factories because potatoes with greater than 0.1% sugar causes darkening or blackening of the end products (chips) (Van der Plas, 1987).

8.4.8 BREAKING OF DORMANCY

The dormancy period of certain roots and tubers are increased under suitable storage conditions after postharvest. Breaking of the dormancy is undesirable as the shelf life is reduced. Some examples of this condition are the growth of sprouts in potatoes, onions, carrots, etc. Sometimes roots are also developed under high moisture conditions. Growth of roots or sprouts decreases the market value of these products (Schouten, 1987). Also, owing to the breaking of dormancy, stored starch is converted into sugars needed for the growing points.

8.5 CONCLUSION

F&V are quite sensitive to many factors, which causes deterioration soon after harvest. In order to increase the keeping quality and life of those harvested agricultural produce, storage plays an important role. Storage of F&V are done for various reasons like to accumulate the produce, transport transit to market/ processing plant, for maintaining regular supply and uniform price round the year, etc. Storage conditions for agricultural produce vary for different commodities to be stored. Few factors affecting the shelf life of harvested F&V are biological factors (microorganisms, insects, rodents), physical factors (temperature, RH), chemical factors (moisture, oxygen), physiological factors (enzymes, respiration), mechanical factors (mechanical damage), etc. By eliminating or reducing the effects of these factors, the storage condition for F&V can be enhanced. Conventional storage facilities for storage of F&V are low temperature storage, controlled atmospheric storage, hypobaric storage, zero energy storage structure, etc. Physiochemical changes which occurs after postharvest and leads to quality loss and deterioration, also differs for different F&V. Chemical composition of F&V varies both within and between species, however, the factors which cause postharvest losses, are considered as the same. A detailed and systematic study of the biochemical reactions and their control is necessary to maximize quality and minimize losses of fresh F&V. Henceforth the future research scope in the domain of fruits vegetable storage lies to have a better understanding about the occurrence of different physicochemical changes and developing economical, safe, and effective methodologies to prevent the same. Effective storage concept would allow storing freshly harvested foods for longer duration of time, without any effects of the quality of the stored product.

ACKNOWLEDGMENT

The authors are grateful to the Centre for Incubation, Innovation, Research, and Consultancy, Jyothy Institute of Technology (Bengaluru, Karnataka) and Indian Institute of Technology, Kharagpur (West Bengal, India) for providing necessary infrastructure, technical help, and expert opinions to complete this work.

KEYWORDS

- **abscisic acid**
- **agricultural market information**
- **engineering design aspects**
- **gibberellins**
- **indoleacetic acid**
- **physicochemical changes**
- **storage structures**

REFERENCES

Abdel-Rahman, M., Thomas, T. H., Doss, G. J., & Howell, L., (1975). Changes in endogenous plant hormones in cherry tomato fruits during development and maturation. *Physiologia Plantarum, 34*, 39–43.

Acked, J., (2002). Maintaining the postharvest quality of fruits and vegetables. In: Wim J., (ed.), *Fruit and Vegetable Processing: Improving Quality* (pp. 119–146). Cambridge, Woodhead Publishing Limited.

APEDA, (2018). *Fresh Fruits and Vegetables*. Agricultural and Processed Food Products Export Development Authority. URL: http://apeda.gov.in/apedawebsite/six_head_product/FFV.htm (accessed on 19 December 2020).

Ben-Yehoshua, S., & Rodov, V., (2003). Transpiration and water stress. In: Bartz, J. A., & Brecht, J. K., (eds.), *Postharvest Physiology and Pathology of Vegetables* (Vol. 2, pp. 111–159).

Ben-Yehoshua, S., Burg, S. P., & Young, R., (1985). Resistance of citrus fruit to mass transport of water vapor and other gases. *Plant Physiology, 79*(4), 1048–1053.

Burden, J., & Wills, R. B. H., (1989). *Prevention of Post-Harvest Food Losses: Fruits, Vegetables, and Root Crops-a Training Manual*. FAO-Food and Agriculture Organization, URL: http://www.fao.org/docrep/T0073E/T0073E00.htm#Contents (accessed on 19 December 2020).

Burton, W. G., (1982). *Post-Harvest Physiology of Food Crops* (p. 339). Longman Group Ltd. New York.

Chakraverty, A., (2005). *Postharvest Technology of Cereals, Pulses, and Oil Seeds*. Oxford & IBH Publication Co. Pvt. Ltd., New Delhi (India).

Chen, H., Yang, H., Gao, H., Long, J., Tao, F., Fang, X., & Jiang, Y., (2013). Effect of hypobaric storage on quality, antioxidant enzyme and antioxidant capability of the Chinese bayberry fruits. *Chemistry Central Journal, 7*(1), 4.

Dadhich, S. M., Dadhich, H., & Verma, R., (2008). Comparative study on storage of fruits and vegetables in evaporative cool chamber and in ambient. *International Journal of Food Engineering, 4*(1).

Dostal, H. C., & Leopold, A. C., (1967). Gibberellins delays ripening of tomatoes. *Science, 158,* 1579–1580.

El-Ramady, H. R., Domokos-Szabolcsy, É., Abdalla, N. A., Taha, H. S., & Fári, M., (2015). Postharvest management of fruits and vegetables storage. In: *Sustainable Agriculture Reviews*. Springer International Publishing Switzerland.

FAO, (2003). *Food Security: Concepts and Measurement*. Food and Agriculture Organization of The United Nations Trade Reforms and Food Security: Conceptualizing the Linkages. URL: http://www.fao.org/docrep/005/y4671e/y4671e06.htm#fn22 (accessed on 19 December 2020).

FAO, (2004). *Manual for the Preparation and Sale of Fruits and Vegetables: From Field to Market*. FAO agricultural services bulletin no. 151. Food and Agriculture Organization of the United Nations, Rome.

Femenia, A., Sanchez, E. S., & Rossello, C., (1998). Effects of processing on the cell wall composition of fruits and vegetables. *Recent Res. Developments in Nutrition Res., 2*, 35–46.

FFTC, (2018). *Postharvest Losses of Fruit and Vegetables in Asia*. Food and Fertilizer Technology Center (Taiwan). URL: http://www.fftc.agnet.org/library.php?func=view&id=20110630151214 (accessed on 19 December 2020).

Frenkel, C., & Eskin, M., (1977). Ethylene evolution as related to changes in hydroperoxides in ripening tomato fruit. *HortScience, 12*, 552–553.

Gast, K. L. B., (2001). *Storage Conditions: Fruits and Vegetables*. Bulletin #4135, Cooperative Extension Publications, University of Maine, Orono, Maine 04469, URL: https://extension.umaine.edu/publications/4135e/ (accessed on 19 December 2020).

Godfray, H. C. J., Beddington, J. R., Crute, I. R., Haddad, L., Lawrence, D., Muir, J. F., Pretty, J. et al., (2010). Food security: The challenge of feeding 9 billion people. *Science, 327*(5967), 812–818.

Hashmi, M. S., East, A. R., Palmer, J. S., & Heyes, J. A., (2016). Hypobaric treatments of strawberries: A step towards commercial application. *Scientia Horticulturae, 198*, 407–413.
IARI, (2018). *Low-Cost Storage Technologies for Preservation of Horticultural Produce and Food Grains*. Indian Agricultural Research Institute- Indian Council of Agricultural Research, URL: https://www.iari.res.in/index.php?option=com_content&view=article&id=195&Itemid=1106 (accessed on 19 December 2020).
ICAR, (2012). *Fruits and Vegetables, Handbook of Agricultural Engineering* (Vol. 37, pp. 529–558). Directorate of Knowledge Management in Agriculture, Indian Council of Agricultural Research (ICAR).
IFPRI, (2018). Food Security, International Food Policy Research Institute (US), URL: http://www.ifpri.org/topic/food-security (accessed on 19 December 2020).
Jay, J. M., (1992). *Modern Food Microbiology*. Aspen Publishers, Inc., Gaithersburg, Maryland.
Jiménez, A., Guillén, R., Fernández-Bolaños, J., & Heredia, A., (1997). Factors affecting the "Spanish green olive" process: Their influence on final texture and industrial losses. *Journal of Agricultural and Food Chemistry, 45*(10), 4065–4070.
Kader A. A., (1983). Postharvest quality maintenance of fruits and vegetables in developing countries. In: Lieberman, M., (ed.), *Post-Harvest Physiology and Crop Preservation* (Vol. 46). Nato Advanced Study Institutes Series (Series A: Life Sciences), Springer, Boston, MA.
Kader, A. A., & Rolle, R. S., (2004). *The Role of Post-Harvest Management in Assuring the Quality and Safety of Horticultural Produce* (Vol. 152). Food & Agriculture Organization.
Lawton, A. R., (1991). *Measurement of Ethylene Gas Prior to and During Transport*. 19th International Congress of Refrigeration, IIR/IIF, Montreal, 1–11.
Li, W., Zhang, M., & Yu, H. Q., (2006). Study on hypobaric storage of green asparagus. *Journal of Food Engineering, 73*(3), 225–230.
Ludford, P. M., (1987). Postharvest hormone changes in vegetables and fruit. In: Davies, P. J., (ed.), *Plant Hormones and Their Role in Plant Growth and Development*. Springer, Dordrecht.
Martínez-Romero, D., Bailén, G., Serrano, M., Guillén, F., Valverde, J. M., Zapata, P., Castillo, S., & Valero, D., (2007). Tools to maintain postharvest fruit and vegetable quality through the inhibition of ethylene action: A review. *Critical Reviews in Food Science and Nutrition, 47*(6), 543–560.
Maxwell, S., & Smith, M., (1992). Household food security; a conceptual review. In: Maxwell, S., & Frankenberger, T. R., (eds.), *Household Food Security: Concepts, Indicators, Measurements: A Technical Review*. New York and Rome: UNICEF and IFAD.
McGlasson, W. B., (1978). Role of hormones in ripening and senescence. In: Hultin, H. O., & Milner, M., (eds.), *Postharvest Biology and Biotechnology* (pp. 77–96). Food and Nutrition Press, Inc., Westport, CT.
Morton, I. D., & Macleod, A. J., (1990). *Food Flavors, in the Flavor of Fruits* (p. 360). Amsterdam, Elsevier, Part C.
Nandy, S. K. (2018). Enzyme use and production in industrial biotechnology. *Research Advancements in Pharmaceutical, Nutritional, and Industrial Enzymology*, Bharati, S. L. & Chaurasia, P. K., (eds.). Published by IGI Global, Medical Information Science Reference (USA), pp. 341.
Potter, N. N., & Hotchkiss, J. H., (2012). *Food Science*. Springer Science and Business Media.
Reid, M. S., (1987). Ethylene in postharvest technology. In: Kader, A. A., (ed.), *Postharvest Technology of Horticultural Crops and Development* (Vol. 13, pp. 257–278). Martinus Nijhoff Publishers.

Rhodes, M. J. C., (1980). The maturation and ripening of fruits. In: Thimann, K. V., (ed.), *Senescence in Plants* (pp. 157–205). CRC Press, Inc., Florida (USA).

Roy, S. K., & Pal, R. K., (1989). A low cost zero-energy cool chamber for short-term storage of mango. In: *III International Mango Symposium* (Vol. 291, pp. 519–524).

Sastry, S., Buffington, D., & Baird, C. D., (1977). Transpiration rates of certain fruits and vegetables. *ASHRAE Journal-American Society of Heating Refrigerating and Air-Conditioning Engineers, 19*(12), 36.

Schouten, S. P., (1987). Bulbs and tubers. In: Weichmann, J., (ed.), *Postharvest Physiology of Vegetables* (Vol. 31, pp. 555–581). New York, Marcel Dekker.

Shakuntala, M. N., & Shadaksharaswamy, M., (2001). *Food: Facts and Principles*. Mohinder Singh Sejwal Publishers, New Delhi/India for Wiley Eastern Limited.

Silva, E., (2010). *Respiration and Ethylene and Their Relationship to Postharvest Handling, Extension-Issues, Innovation, Impact: A Part of Cooperative Extension System*. URL: http://articles.extension.org/pages/18354/respiration-and-ethylene-and-their-relationship-to-postharvest-handling (accessed on 19 December 2020).

Srivastava, R. P., & Kumar, S., (1994). *Fruit and Vegetable Preservation: Principles and Practices*. International Book Distribution Company, CBS Publishers & Distributors Pvt. Limited, Uttar Pradesh (India).

Thompson, W. W., Lewis, L. N., & Coggins, C. W., (1967). The reversion of chromoplasts to chloroplasts in Valencia oranges. *Cytologia, 32*, 117–124.

Tsay, L. M., Mizuno, S., & Kozukue, N., (1984). Changes of respiration, ethylene evolution, and abscisic acid content during ripening and senescence of fruits picked at young and mature stage. *Journal of the Japanese Society for Horticultural Science, 52*, 458–463.

Tucker, G. A., (1993). In: Seymour, G. B., Taylor, J. E., & Tucker, G. A., (ed.), *Introduction, Biochemistry of Fruit Ripening* (pp. 53–81). Chapman and Hall, London.

Van, D. P. L. H. W., (1987). Potato tuber storage: Biochemical and physiological changes. In: *Potato* (pp. 109–135). Springer, Berlin, Heidelberg.

Van, D. V. H. C. P. M., & Donkers, J. W., (1994). Physiological and biochemical changes in cell walls of apple and tomato during ripening. In: 6th *International Symposium of the European Concerted Action Program* (pp. 19–22).

Varga, A., & Bruinsma, J., (1974). The growth and ripening of tomato fruit at different levels of endogenous cytokinins. *Journal of Horticultural Science, 49*, 135–142.

CHAPTER 9

CHARACTERIZATION OF MODERN COLD STORAGE FOR HORTICULTURE CROPS

JINKU BORA, THOITHOI TONGBRAM, MIFFTHA YASEEN, MUNEEB MALIK, and ENTESAR HANAN

Department of Food Technology, Jamia Hamdard, New Delhi–110062, India, E-mail: jinkubora@gmail.com (J. Bora)

ABSTRACT

Cold storage is at the heart of horticulture harvest, transportation, storage, marketing and consumption. In an attempt to minimize physiological, microbial and transportation losses, low temperature cold storage or refrigerated storage techniques or a combination of both are used right after harvest. The type of techniques or the sequential combination of techniques used will depend upon the nature of harvest, packaging, capacity, efficiency, resources and utility of produce. For an efficient cold storage system, the location of cold storage facilities must be at strategic points for ease of access, distribution, commercialization, supply and maintenance. Proper planning, attention to temperature and relative humidity requirements of stored produce, produce management and hygiene management provide further reinforcements in making the entire cold storage chain function competently. In supplementation to cold storage, controlled atmosphere storage and modified atmosphere storage are necessary to maintain uniformity and control over storage environments. Sometimes, the need to store different produce arises, and in such cases, utmost care is demanded while operating mixed commodity storage. Failure to maintain necessary storage conditions due to unforeseen circumstances or fluctuations in power supply may lead to disorders such as chilling injury and freezing injury, which may bear heavy produce and capital loses. For small scale systems, farmers and entrepreneurs, alternative techniques like surface cooling and evaporative cooling, aided by refrigerated transport systems can be used as gap-fillers to maintain continuity in cold supply chain. The ultimate aim of modern cold storage systems largely encompass, but not limited to, reducing post-harvest losses and bridging the gap between exploding demand and variable supply of produce. Therefore, planning, evaluation and effective implementation of such systems is a matter of global necessity.

9.1 INTRODUCTION

Horticulture crops encompass the science and art of growing plants like fruits, vegetables, flowers, and any other cultivars. Most of the fresh fruits and vegetables (F&V) can be grouped under the perishable type of food. With the exception of a few (e.g., onions), these commodities are likely to get spoiled or decayed, even become unsafe to consume within a few days, if not stored properly under ambient temperature conditions. A study suggests that roughly one-third of food produced for human consumption is lost or wasted globally, which amounts to about 1.3 billion tons per year (Gustavsson et al., 2011).

Cold storage units in the supply chain can provide for an effective as well as an affordable means of storing F&V at the commercial level. The use of cold storage units is justifiable because a cost-effective cooling in any form during storage is better than storage without any cooling at all. Its effectiveness is manifested in the marked reduction of the post-harvest losses.

According to the data provided by the AMI of Bonn, the global production of F&V has reached 1.74 billion tons in 2013 (9.4% increase from 1.59 billion tons in 2012). Of which 950 million tons of were vegetables and 790 million tons were fruits (WFO, 2014). An IARW global cold storage capacity report (2014) states that the total capacity of refrigerated warehouses is estimated at 552 million cubic meters worldwide in 2014, with an increase of 92 million cubic meters (20%) over 2012. The global refrigerated warehouse storage capacity is estimated to be growing at a rate above 5% per year during a sustained period of time in 17 countries (Salin, 2014).

The ever-growing population of the world presses that the production of F&V be increased every year in order to bridge the gap between the supply and the demands. Improved irrigation, use of improved varieties, or advanced methods of cultivation and harvesting are alone not enough to feed the world. Thus, the need for an effective system of cold storage units for F&V at various points in the supply and distribution chain after harvest is a must. This chapter sheds limelight into the importance, scope, and functioning of a commercial cold storage unit with focus on F&V.

9.2 PRINCIPLE OF COLD STORAGE

We must keep in mind that F&V respire and the idea of cold storage is to keep the overall respiration rate to a minimum low as possible. In addition, stored commodities can eventually become a site for microbial activity and hence, spoilage, decay, and deterioration. The objective is to keep the stored produce to a temperature as minimum as possible such as to suppress any metabolic, physiological, or any microbial activity that may be associated with the 'living tissues.'

Since for every 10° rise or fall in the temperature, there is a corresponding increase or decrease in the respiration rate, a carefully controlled and maintained low-temperature environment of the stored produce is a necessity to halt any deteriorative physiological changes (respiration, ethylene production, enzymatic reactions, water loss, etc.). The principle of cold storage units, thus, lies herein the fact that a reduced temperature environment can increase the shelf-life of the produce by keeping in check the various activities detrimental to the storage life of the produce.

Cold storage at the commercial level can be carried out in the form of cooling storage techniques or in the form of refrigerated storage techniques.

9.2.1 TYPES OF COLD STORAGE

9.2.1.1 LOW-TEMPERATURE COLD STORAGE

9.2.1.1.1 Pre-Cooling Techniques

The residual heat or the field heat associated with any fresh harvest requires that heat be immediately removed by cooling techniques. The rapid cooling may be provided indirectly by circulating cold air or directly by the use of cold water or ice. This process of rapid removal of field heat or residual heat from the produce directly after harvest is known as pre-cooling.

The ambient temperatures to which the produce must be brought down to, however, depend on the chilling sensitivity of each item. Cooling reduces metabolic activities and increase shelf life. It also retards microbial activities and arrests pathogen proliferation.

Pre-cooling serves as one of the most effective means of reducing post-harvest losses, stating both in terms of efficiency and cost-effectiveness. In fact, it is the first step to any type of cold storage, particularly for produce such as F&V that are perishable in nature if the storage temperatures are not brought down to a controlled level. By removing some of the heat of the produce, pre-cooling helps lower down the energy requirements during storage cooling by refrigeration.

From a commercial point of view, the following factors are to be taken into account before implementation of any type of pre-cooling method:

- **Nature and Type of Produce:** Different produces (tropical, subtropical, temperate, etc.), require different storage temperatures with temperate produces requiring near-freezing temperatures which are injurious to tropical and sub-tropical produces. The produce temperature needs to be brought down according to its ambient storage temperature.
- **Design and Type of Packaging:** The type of material used for packaging, the size and the number of holes on the packaging as well as the mode of palletization affects the rate at which the heat can be removed from the produce.
- **Cooling Capacity:** Different methods of cooling have varying capacities, and the time taken to cool a particular volume of produce is determined by the production capacity. Methods with faster rates of cooling are to be used to cool large volumes of produce.
- **Economic Efficiency:** The cost of construction, operation, and maintenance are to be strictly considered along with electricity cost. It is not wise to use a high cost production line for a low produce volume as it will lead to higher prices at the consumers end, and the cost of cooling cannot be justified by the profit margin from the end-selling price.
- **Resource Availability:** Resources such as land, electricity, manpower, transportation, and the social income status of the consumers are to be taken into account before the establishment of a cooling infrastructure. Low cost and simple cooling methods for low-income areas would prove to be as much an efficient operation just as those that put in more cost and resources for the upper-income strata.

Although the objective and principle of pre-cooling remains the same, different methods are used to justify the cooling purpose and cater to end-consumer requirements. There are seven principal methods:

- room cooling;
- forced-air cooling;

- hydro-cooling;
- ice cooling;
- vacuum cooling;
- cryogenic cooling; and
- evaporative cooling.

9.2.1.1.2 Room Cooling

It is one of the oldest, well-established, cheap, and affordable methods of pre-cooling. Room cooling is usually done in a refrigerated room and is important for produces that are sensitive to free and surface moisture. This method involves placing the produce in wooden, fiberboard or plastic boxes, which are then exposed to cold air inside a cold room. The well-vented packages are properly stacked to allow the cold air to pass through and inside the packages.

A typical room cooling storage has discharge vents at the ceiling that releases cold air into the room, the cold air passing through the produce and returning back to the heat exchanger. The cooled air is generally supplied by forced or induced draft coolers, consisting of framed, closely spaced and finned evaporator coils fitted with fans to circulate the air over the coils.

To achieve adequate cooling, air velocities of 60 m/min and relative humidity (RH) of 90% to 95% is to be maintained. Spacing (between the containers and walls: 6 to 12 inches and between the boxes and ceiling: 18 to 24 inches) must be maintained and the produce should be cooled down within 24 hours of its harvest.

Room cooling is not suitable for produce cooling in large bins or pallets as it tends to be able to cool the produce only at the periphery while at the center, respiration generates more heat than that can be removed by cooling. If done properly in small volumes, a properly designed room cooling can be relatively energy efficient. Because of its slow nature of cooling and small capacity, the applicability of this method to large-scale commercial cold storage is, however, limited. Room cooling is inadequate for produce such as strawberries that require immediate cooling after harvest.

9.2.1.1.3 Forced-Air Cooling

This is the most commonly used method of pre-cooling. It is 75% to 90% greater energy-efficient than room cooling (Kanlayanarat, Rolle, and Acedo, 2009). It is the same as room cooling, or rather a modification of it; only that it uses fans that pull cold air through the packed produce, thus increasing the heat transfer and removing the heat from the produce faster in the process. The rate of cooling depends on the temperature of the cold air and the airflow rate.

Forced air-cooling is used for produce that demand a faster and immediate removal of heat after harvest. For a successful application, the vented packages are to be stacked in a definite pattern and the vents facing the direction of the moving cold air. The cold air moves inside the package rather than around it (in case of room cooling) and serves as an efficient heat transfer mechanism.

Commercially, different forced air cooling arrangements are available, and depending upon produce suitability, forced-air cooling can be done by: (a) high velocity air circulated inside a cold room; (b) pressure gradients on between two sides of the packed produce, thereby creating airflow; (c) continuous conveyors on which cold air is forced through the void spaces of the produce in bulk.

The only factor that governs forced air-cooling is the air flow rate and the minimum temperature of the air to avoid chilling injuries. In general, the cool air necessary for this type of cooling can be generated from (a) direct expansion refrigeration system; (b) ice bank cooling system; and (c) water cascade. Forced air-cooling uses blowers (centrifugal fans, also known as squirrel cage fans) or axial fans to push cold air through the produce. The choice of the fans depend upon the type and the quantity of produce to be cooled, the static pressure involved (axial fans for less than 4 in. water and centrifugal fans for greater than 4 in. water) and also the arrangement of the packed produce (bulk, box or stacked). Differential pressures of approximately 0.6 to 7.5 mbar and airflow rates of 0.001 to 0.003 m^3/s kg produce are generally in use.

9.2.1.1.4 Hydro-Cooling

Hydro cooling is used for produce that are not sensitive to wetting (e.g., cucumber, orange, green onions, etc.). The commodity as well its packaging materials must be tolerant of wetting, chlorine (used to sanitize the hydro-cooling water) and water beating damage (Mitchell et al., 1972). At favorable flow rates and temperature gradients, water removes heat approximately 15 times faster than air, and hence this form of pre-cooling is faster than room cooling or forced-air cooling. However, it is only about 20% to 40% energy efficient as compared to 70% to 80% for room cooling or forced air cooling (forced-air cooling more efficient than room cooling) (Kanlayanarat et al., 2009).

Cooling involves flooding; spraying or immersing the produce in or with water and so proper sanitation of water must be maintained. Use of mild disinfectants such as chlorine at concentrations of 100 ppm (measured as hypochlorous acid) is recommended to avoid contamination and decay hazards with re-circulated water. For effective cooling, water should be allowed to enter the packages, whether palletized or packed in wooden crates, waxed fiberboard cartoons, mesh poly bags, or bulk bins. Once cooled, the produce temperature must be kept maintained.

Hydro-coolers can be of the flood type, the batch type, and the immersion type. The conventional flood type hydro-cooler floods the produce conveyed in a cooling tunnel while the batch type sprays water over the produce over a length of time decided by the incoming produce temperature and the environmental conditions (season). The batch type can only cool a small amount of produce as compared to the conventional type and therefore is less expensive.

For a cooling process to be efficient, the contact between the water and the produce must be uniform. However, this cannot be achieved by either of the two methods. The immersion type overcomes this problem by a combination of both methods. The produce first gets immersed in cold water (immersion type) until a conveyor carries them out and though an overhead shower of cold water (batch type). This allows the produce to be surrounded by chilling water and the greater contact surface leads to a faster heat removal rate and hence, a rapid cooling of produce nearly twice as fast as the conventional and the batch types is possible. In addition, it allows greater packaging flexibility (cooling followed by immediate packaging) and is suitable to a wide range of growers.

Hydro cooling proves to be advantageous over other pre-cooling techniques in that it prevents moisture loss during cooling, and it is comparatively very rapid. Field heat can be removed in a matter of 20 to 30 minutes which otherwise takes hours if done by forced-air cooling. A combination of hydro and air-cooling, hydro-air-cooling, in which a fine mist of refrigerated air and water is circulated through the stacks of produce, is gaining importance by reducing water requirement and thus creating a potential for improved sanitation. This method is influenced by the ratio of air-water that determines the heat transfer capability and its applicability to different produces.

9.2.1.1.5 Ice Cooling

Before the advent of the modern techniques of pre-cooling, cooling with the use of crushed or fine granular ice packed around produce in cartons or sacks, or ice slurry injected into waxed carton packed produce was extensively used to pre-cool produce and maintain temperature during transit. This method of pre-cooling is known as ice cooling or contact or package icing. Ice cooling has its comparative advantages over other methods of cooling. Not only ice makes the rapid heat transfer possible on application, but it also removes heat from the produce/package as it melts. Because of this residual effect, ice cooling works well with produce that have high respiration rates. Due to the increasing availability of ice, icing is being increasingly used in developing countries. Different varieties of methods have been employed for applying ice to packed produce such as to achieve the required cooling.

Individual package icing is the simplest method of icing in which a measured amount of ice, whether slush, flaked, or crushed, is manually placed over the packaged produce. Although successful in many cases, the cooling is uneven as the ice usually tends to stay where it has been placed until it has melted. In addition, individual package icing requires that each individual carton be opened, iced, and packed, and the entire process is comparatively slow. Due to this, this technique may not be recommended for high volume production. However, with the use of ice dispensing machines, packaging conveyors and roller benches have automated the process up to an extent.

In liquid icing, a slurry of crushed ice and water is used instead of plain crushed ice to sustain better cooling requirements. The ice slurry is either pumped into open containers or if packed/palletized, it is injected inside the package through vents or handholds without de-palletizing. Liquid icing is an excellent cooling method in that the contact between the ice and the produce is greatly improved as the slurry easily distributes inside the package and in between the produce. It can be considered a compromise between package icing and hydro cooling. A downside might be the possibility of a site for microbial contaminations as the produce is wet and often warm.

Top icing is another method of icing in which ice is manually or mechanically placed on top of packed containers, more often to supplement other cooling. Wooden crates have been replace by CFB and then by waxed impregnated corrugated containers that have allowed the use of top icing. Ice bottles wrapped with papers are used for cooling fresh produce, particularly used to cool high-value produce during transportation in China. The paper wrap prevents direct contact between the ice bottles and the produce.

The advantages of ice cooling can be laid out as follows:

- Icing is relatively energy efficient. One pound of ice will cool as much as three times its weight from 29.4 to 4.4°C (Kanlayanarat et al., 2009).

- The produce is rapidly cooled and remains to be cooled even as the ice is melting. In addition, because of the melting ice, the cooled produce does not get dry.
- In addition to removing field heat, it can continue to maintain low temperatures for short transits and thus eliminating the need for refrigeration equipments during transportation.

While the use of ice cooling offers interesting advantages, several ambiguities exist in its application. The additional weight of the ice increases costs, decreases volume inside the package and the package is left only partly full when the ice has melted. Another disadvantage may be that of wetting produce due to melting ice creating a warm wet produce (a site for post-harvest diseases and soft rots) although not allowing the produce to re-warm once iced might solve this. The ice must be free of physical, chemical, and biological hazards to avoid undue contamination. As for the packaging material, any container that can withstand wetting for long periods without any compromise on the strength can be used. Waxed fiberboards, perforated plastic liners, wooden bound wire crates and baskets are popularly used, waxed fiberboard being best suited for icing operations.

9.2.1.1.6 Vacuum Cooling

This technique is effective with a high surface to volume ratio and for produce which is very difficult to cool with other techniques that uses cold air or water. Vacuum cooling is based on two principles-at reduce pressure, water evaporates at a lower temperature; and when water evaporates, it absorbs some heat known as latent heat of vaporization. The produce is kept inside a vacuum chamber such as a large metal cylinder with much of the air evacuated.

At reduced pressures (say 4.6 mm Hg, the normal atmospheric pressure being 760 mm Hg), the water in the produce starts to evaporate. In doing so, it takes up heat from the produce, thereby reducing the temperature as rapid as in about 10 to 30 minutes. The weight loss is accounted at about 1% for every 5 minutes making the average weight loss to about 1.5% to 5%. To avoid excessive water loss, water in the form of fine mist can be applied during the vacuum process. The technique is being currently popular in Taiwan and Japan.

9.2.1.1.7 Cryogenic Cooling

This technique uses the latent heat of vaporization of liquid nitrogen (LIN) or solid carbon dioxide (dry ice). LIN evaporates at –196°C while solid carbon dioxide vaporizes at –78°C. This forms the basis of cryogenic pre-cooling. The produce is conveyed through a tunnel where LIN or solid carbon dioxide is sprayed or applied and is evaporated at controlled rates. The evaporation step is critical as to not freeze the produce but cool it by maintaining careful control of the conveyor speed and the evaporation rate.

Cryogenic cooling is comparatively cheaper to install, but the operation and maintenance can be expensive. Its main application lies with seasonal crops such as soft fruits and thus, justifies its cost over such period of use, which otherwise would be associated with high capital cost when other cooling techniques be used. Due to the high cost and limited availability of LIN, dry ice or other non-toxic refrigerants, this technique is more suitable for use with relatively expensive products.

9.2.1.1.8 Evaporative Cooling

Evaporative cooling is simply based on the principle of latent heat of vaporization of water. It is an inexpensive technique, and it is effective in regions where the RH of the surrounding air is low. The produce is stacked in vented containers, and dry air is drawn through a moist padding or a fine mist shower and through the vented containers of produce with a suction fan/pump. As the air saturated with the fine mist pass through the produce in the containers, the mist deposits on the surface of the produce. In the process of conversion of liquid water to vapor, heat is removed from the produce, thereby lowering the temperature. A RH of less than 65% is required for effective cooling. This technique is suitable for warm-season crops with warmer storage temperatures (7.2 to 7.7°C).

Evaporative coolers can be constructed to cool the air in an entire storage structure or just a few containers of produce. These coolers are best suited to lower humidity regions, since the degree of cooling is limited to 1 to 2°C (2 to 4°F) above the wet-bulb temperature. A cooling pad of wood fiber or straw is moistened, and air is pulled through the pad using a small fan. In the example provided here, 0.5 gallons of water per minute is dripped onto an 8 square foot pad, providing enough moist air to cool up to 18 crates of produce in 1 to 2 hours. Water is collected in a tray at the base of the unit and re-circulated (Kitinoja and Kader, 1995).

9.2.1.2 REFRIGERATED COLD STORAGE

Cold storage or refrigerated storage is by far the most effective method of preserving quality and extending shelf life of produce such as F&V. Cold storage slows the biological activity of produce, growth of detrimental/deteriorative microorganisms, reduces the rate of moisture loss from the produce surface, and decreases susceptibility to ethylene gas damage. Inside the cold storage, a balance between produces requirements of temperature and RH must be maintained to ensure effectiveness. Tropical produces demand higher temperatures of storage beyond 10°C while sub-tropical and temperate produces are suitable for storage at lower temperatures approaching 0°C. A nearly saturated RH of 85% to 95% is needed to maintain moisture content in the produce, for uniformity in ripening, to reduce instances of decay development and pathological conditions.

It is a matter of profitability that a producer will always try to sell off produces at cheap rates immediately after harvest due to fears of losses during bulk storage. However, a producer that has access to cold storage has greater flexibility before he decides to shift his produces to the market. A strategically located cold storage close to the field of production can help in post-harvest management and transportation. With the combination of appropriate pre-cooling techniques and cold storage facilities, post-harvest losses during storage, transportation, distribution, and marketing and also in transit between each point can be reduced considerably thus, ensuring a greater volume of produce that can be available to the consumers.

9.2.1.2.1 Refrigeration System

For use in the food industry, a refrigeration system either uses closed mechanical refrigeration cycles that circulates refrigerants or open cryogenic systems with cryogens such as LIN and solid/ liquid carbon dioxide (CO_2).

9.2.1.2.2 Closed Mechanical Refrigeration System

A closed mechanical refrigeration system (Figure 9.1) consists of the following components-condenser, compressor, evaporator, and expansion valve. Some commonly used refrigerants in closed mechanical refrigeration system are hydrochlorofluorocarbon (HCFC) and ammonia (Barbosa-Cánovas et al., 2005).

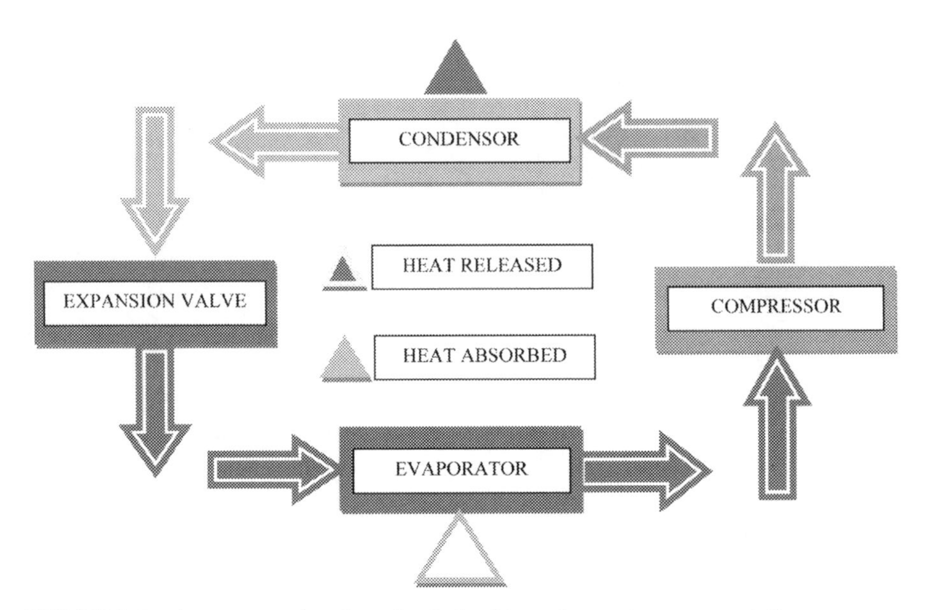

FIGURE 9.1 A one-stage closed mechanical refrigeration system representation.

The process of refrigeration is basically a change of phase of the refrigerant in a closed system with change in pressure and temperature. The compressor sucks in the refrigerant in the vapor phase from the evaporator. Inside the compressor, high pressure, leads to a decrease in volume of the refrigerant and the temperature rises as it gets compressed. The compressed vapor then passes to the condenser to release heat which causes the vapor to be converted to a liquid state. The liquid refrigerant flows into the expansion valve at reduced pressure to expand and become a liquid-gas mixture. This mixture is fed into the evaporator housing the produce to be frozen and it takes up heat from it to become a saturated gas phase. Heat is thus removed from the produce, and the vapor phase is again sucked in inside the compressor for another cycle of freezing.

9.2.1.2.3 Open Cryogenic Systems

The open cryogenic system employs LIN or solid/liquid CO_2. Here, unlike in a closed system, the cryogen is not recirculated and is consumed differently than in the closed mechanical system. In such traditional cooling systems, the spent vapors are expelled into the atmosphere and not recovered for reuse.

9.2.1.2.4 Refrigerants

The choice of refrigerants should be decided based on the physical, thermodynamic, and chemical properties of the fluid (Barbosa-Cánovas et al., 2005). In addition, economic, environmental, and hazardous nature of the fluid should also be in question. For commercial uses in food industries, ammonia followed by chlorofluoromethane and tetrafluoroethane are mostly used while halocarbons have been banned for potential hazardous effects (Stoecker and Jones, 1982; Persson and Londahl, 1993).

9.2.1.3 COMMERCIAL COLD STORAGE FACILITIES

Commercial cold storage facilities can be located at crucial points-packing houses, collection points, distribution points, trucking houses, dry ports or in locations close to airports or export harbors (Kanlayanarat, Rolle, and Acedo, 2009). While selecting a location for set up, the following considerations must be fulfilled:

- **Accessibility:** The site must be easily accessible or connected by road or by other means.
- **Location:** The site must be situated such that it offers an unidirectional flow of produce from the field towards other destinations through the cold storage site;
- **Power Supply:** Not only the capacity of the cold storage is determined by the availability of power supply at the area but back up supply lines/generators are also a must.
- **Water Supply:** Sanitization of the facility with good quality water free from physical, chemical, and biological hazards.
- **Site Drainage:** Good drainage is a necessity to remove out waste and condensates.
- **Waste Treatment/Disposal:** Waste treatment, disposal, and management must be adequate and according to plant capacity.

The capacity and design of the storage facility depends upon:

- Produce volume to be stored;
- Types of containers to be used in storing produce;
- Volume occupied per container;
- Mode of operation whether long term or for transit;
- Aisle, lateral, and headspace needed;

- Available space from the site.

9.3 TEMPERATURES AND RH FOR DIFFERENT PRODUCTS (SOURCE: MCGREGOR, 1989)

9.3.1 FRUITS AND VEGETABLES (F&V), 0 TO 2°C (32 TO 36°F), 65–75% RELATIVE HUMIDITY (RH)

Moisture will damage these products: garlic, onions (dry).

9.3.2 FRUITS AND VEGETABLES (F&V), 4.5°C (40°F), 90–95% RELATIVE HUMIDITY (RH)

Cactus leaves, cactus pears, caimito, cantaloupes**, clementine, cranberries, kumquat, lemons*, lychees, mandarin*, oranges (California and Arizona), Pepino, tamarillo, tangelos*, tangerines*, ugli fruit*, yucca root [*citrus treated with biphenyl may give odors to other products; **can be top-iced].

9.3.3 FRUITS AND VEGETABLES (F&V), 10°C (50°F), 85–90% RELATIVE HUMIDITY (RH)

Many of these products are sensitive to ethylene. These products also are sensitive to chilling injury. Beans, calamondin, chayote, cucumber, eggplant, haricot vert, kiwano, malanga, okra (sot shell), olive, peppers, potatoes (storage), pummel, squash, summer, tamarind, taro root.

9.3.4 FRUITS AND VEGETABLES (F&V), 13 TO 15°C (55 TO 60°F), 85–90% RELATIVE HUMIDITY (RH)

Many of these products produce ethylene. These products also are sensitive to chilling injury. Atemoya, avocados, babaco, bananas, bitter melon, black sapota, boniato, breadfruit, canister, carambola, cherimoya, coconuts, feijoa, ginger root, granadilla, grapefruit, guava, jaboticaba, jackfruit, langsat, lemons*, limes*, mamey, mangoes, mangosteen, melons (except cantaloupes) (hard shell), papayas, passion fruit, pineapple, plantain, potatoes (new), pumpkin, rambutan, santol, soursop, squash (winter), sugar apple, tomatillos, tomatoes (ripe) [*citrus treated with biphenyl may give odors to other products].

9.3.5 FRUITS AND VEGETABLES (F&V), 18 TO 21°C (65 TO 70°F), 85–90% RELATIVE HUMIDITY (RH)

Jicama, sweet potatoes*, watermelon*, pears, tomatoes, white sapota (for ripening), mature green, yams* [*separate from pears and tomatoes due to ethylene sensitivity].

9.4 TEMPERATURE AND RELATIVE HUMIDITY (RH) MANAGEMENT

1. **Temperature:** A uniform temperature must be maintained within the cold storage chamber. Alternation of warm and cold temperatures can cause produce to sweat by deposition of moisture and can become a site for decay and deterioration. Temperature fluctuations can be maintained by routine manual checks for cold/hot spots at various points in the chamber. Thermometers should be installed for temperature updates and thermostats to control temperature fluctuations as well as for operation of the refrigeration unit. An uniform and unrestricted airflow is a must along with well-insulated and adequate refrigeration capacity. In addition, maintaining the temperature of the refrigeration coils to within 1°C of the air temperature is necessary.
2. **Relative Humidity (RH):** It can be recorded with a hygrometer and kept in check using a humidistat. A drop-in RH of the chamber can be compensated using: (a) water sprays in fine mist form along the air supply line after the refrigeration unit; (b) regulation of exhaust and air circulation; (c) use of moisture-proof barriers in chamber walls and on transit vehicles; and (d) wetting the cold storage chamber with good quality water.

9.5 PRODUCE MANAGEMENT

Produce management comprises of establishing uniformity and control over temperature, RH, air circulation, avoidance of mixing of incompatible produce in the same chamber and spacing needed between stacks/containers for working area and ventilation. A cumulative of these control parameters supported with a good produce inflow and outflow add up to a well maintained cold storage facility. The following points are necessary and should be noted for produce management:

- **FIFO:** Produce must be sorted out on a FIFO (first in first out) basis. Only marketable produce must be sorted out and kept in storage.
- **Pre-Cooling:** This prior to loading in cold storage is a must. Not only will it reduce cooling energy requirements but also avoid large temperature fluctuations inside the chamber upon fresh loading. If pre-cooling facility is not available, the daily intake should be restricted to only about 10% of the storage capacity.
- **Temperature:** The temperature of the cold chamber should be brought down to the temperature appropriate for the produce to be stored about 3 days prior to loading and a week prior if floors of the cold room are not insulated.
- **Stacking:** of containers or packaged produce should ensure proper ventilation for circulation of cold air through and around the packages and throughout the room. A minimum gap of 8 cm between packages and the wall and the floor and 20 cm between drip trays under the cooling evaporators and the top of the stack (headspace).
- **Gaps and Ventilations:** While using containers without vents, gaps of 1 cm between each stack must be ensured and for bins with storage capacities of 200 to 500 kgs, about 8 to 10% of the base floor area should be ventilated with pores/air gaps (Kanlayanarat et al., 2009).
- **Storage Practices:** The storage intake personnel responsible for handling the produce once it arrives must make sure all arrangements are made in advance and in order. The personnel must ensure all documentations are properly completed once the shipment arrives, and a careful examination of the temperature, hygiene, state of the produce upon arrival must be carried out and noted in documents for record-keeping. The documentation should include the anticipated time of arrival, allocated chambers in the cold room, and the temperature at which the produce has to be correspondingly stored. The palletized produce must be placed in pre-assigned locations and the damaged cargo should be kept at the demarcated holding bay.
- **Documentation:** The location of the inventory must also be recorded in the documentation. Inventory records should include the type, quality, and quantity, harvest date, packing date of the produce, pre-cooled

method used, storage entry date, and any other special treatments used on the stored produce.

9.6 HYGIENE MANAGEMENT

Sanitary conditions are to be maintained in cold storage rooms to avoid possible agents of pathogen and contamination build-up on the inside surfaces of the cold room. Cleaning the walls, floor, and other contact surface with a solution of sodium hypochlorite or other non-toxic sanitizers at the end of every season is recommended. While cleaning, the workers should have their personal protection gear on as continuous exposure to high inoculums levels and disinfectants can be a potential health hazard.

Surface mold and fungal spores can be kept in check with the use of ozone (fungicidal) only that ozone is toxic to humans. Personnel working 8 hours per day can be exposed to levels of about 0.1 ppm, and the limit of exposure is attained when the smell of ozone becomes noticeable. Produce respire even when they are being stored under low physiological activity. This causes a gradual build-up of carbon dioxide, ethylene, and other related off-odors and gases. Carbon dioxide becomes toxic beyond 5000 ppm (0.5%) (Kanlayanarat et al., 2009). Proper ventilation and a complete volume room air change cycle should be made for every 6 hours. An inlet valve placed near the rear end of the low-pressure zone and outlet at the opposite end of the high-pressure zone helps with the air change. However, excess ventilation can bring in unwanted moisture, and an additional defrost cycle will need to be added on.

9.7 SUPPLEMENTS TO COLD STORAGE

9.7.1 CONTROLLED ATMOSPHERE (CA) AND MODIFIED ATMOSPHERE (MA)

Controlled atmosphere (CA) storage provides a gaseous environment in addition to the cold temperature. By striking a balance between the various gaseous constituents (carbon dioxide, oxygen, ethylene, etc.), through the use of gas generators/quenchers, the produce respiration can be tuned down further to a safer level, which otherwise would not be possible by cold application alone. Integration of CA systems to the already existing costly cold storage system will further increase cost, and for commercial purposes, it is vital that the added cost must be compensated in terms of increased produce value and storage longevity.

Modified atmosphere relies on produce metabolism, that is, a saturated environment created by produce respiration overtime inside a sealed package. The process can be of passive nature when left to the produce to gradually create the saturated environment or actively by flushing the package with variable quantities of gases that affect metabolism. It is less precise in the sense that the atmosphere created at the time of packaging is bound to be variable at different times under different conditions.

CA and MA application in cold storage can function only if proper temperature and RH conditions are maintained inside the storage room. It also requires that each of these techniques need to function synchronously to allow the system to slow down produce metabolism and its aging process.

9.7.2 MIXED COMMODITIES STORAGE

Sometimes, the need to store different types of produce arises. Under such cases, care should be taken such that the produces are grouped according to their compatibility in terms of temperature, RH requirements, or ethylene production/sensitivity and odor production/sensitivity:

1. **Temperature:** Temperate/sub-tropical produce that require near-freezing temperatures should never be clubbed with tropical ones that require warmer temperatures. Temperate/sub-tropical produce is sensitive to temperatures above 10°C and storage at such temperature may cause the produce to lose moisture, shrivel, and dry up faster. Tropical produces suffers from chilling injuries if stored in less than 10°C.
2. **Odor Transfer:** Earthy, unpleasant aroma and characteristic odor transfer takes place when produces are stored as mixed loads. Therefore, certain combinations should be avoided to prevent such cross-transfers. They may include storing of apples or pears with cabbage, celery, carrots, potatoes, or onions; celery with onions or carrots, citrus, onions, potatoes, and nuts should be kept separately from other F&V (Kanlayanarat et al., 2009).
3. **Ethylene:** High ethylene producers such as ripe bananas, apples, cantaloupe, etc., can induce physiological disorders and/or undesirable changes in color, flavor, and texture in ethylene sensitive or low ethylene producing commodities such as lettuce, cucumbers, carrots, potatoes, etc., (Kanlayanarat et al., 2009).

9.7.3 DISORDERS

Disorders in cold chain arise from failure to maintain temperatures best suited to the produce under storage. Frequent fluctuations in storage temperature, loss of cooling and re-warming, poor management, chilling, and freezing injuries constitute some of the encountered problems:

1. **Chilling Injury:** Fruit and vegetable crops often are susceptible to chilling injury when cooled below 13 to 16°C (55 to 60°F) (Kitinoja and Kader, 1995). Tropical produces suffer from chilling injuries especially when they are stored below their freezing points, requiring temperatures above 10°C. The degree of the damage depends on the period and extent of exposure. The susceptibility however depends on the stage of maturity at harvest and the degree of produce ripeness. It is important to avoid chilling injury, since symptoms include failure to ripen (bananas and tomatoes), development of pits or sunken areas (oranges, melons, and cucumbers), brown discoloration (avocados, cherimoyas, eggplant), increased susceptibility to decay (cucumbers and beans), and development of off-flavors (tomatoes) (Shewfelt, 1990).
2. **Freezing Injury:** This type of injury is caused due to the presence of soluble solids and the inability to determine the exact freezing point of a produce. This condition is due to the fact that the soluble solids vary not only between different produce items but also between the same commodities and even between different parts of the same produce. Thus, a single temperature, even if maintained properly, will not be able to match the varying freezing points of the commodities in a load. Freezing injury is commonly found in refrigerated transportation vehicles. However, the actual cause is observed to be because of equipment malfunction and the fluctuation in temperature. The produce may appear waterlogged or glassy when it reaches non-freezing temperature due to the phase conversion and redistribution of water inside the produce.

9.7.4 ALTERNATIVE METHODS

Cold storage or refrigerated storage of produce requires large capital in term of both set up, operation, and maintenance. This is specially a burden to small-scale entrepreneurs in developing countries and for small farmers. An alternative to this must be sorted out to avoid any breakage in the cold supply chain from the farm right up to use by consumers.

9.7.4.1 EVAPORATIVE COOLING

Evaporative cooling is inexpensive, easy-to-set-up and of low cost. It can be an useful technique as an alternative means to achieve cooling without the burden of high-capital investments. In addition, since the technique is relatively cheap, a produce cooled using such means can be marketed at a fairly cheap price, thus making it easier even while marketing.

One such example is the zero-energy cool chamber in which water is kept in the vicinity of the produce to be cooled. The heat dissipated from the produce is taken up by the water and gets evaporated. Although the drop in temperatures is minimal as compared to those offered by refrigeration techniques, the increase in RH around the produce up to 90% keeps the produce from moisture loss, shriveling, wilting, ripening, etc.

These cool chambers are able to maintain temperatures at 10 to 15°C below ambient, as well as at a RH of 90%, depending on the season (LalBasediya et al., 2013). During the hot summer months in India, this chamber is reported to maintain an inside temperature between 15 and 18°C (59 and 65 F) and a RH of about 95% (Kitinoja and Kader, 1995). Apart from this, several other evaporative coolers have been designed and used by small farmers for temporary storage of produce.

9.7.4.2 SURFACE COATINGS

Surface coatings act as barriers to moisture loss and help prevent produce decay by shielding produce-to-produce surface contacts. In addition, it can reduce weight loss and provide a modified atmosphere on the produce surface, especially for fresh produce by partial or complete blocking of the stomata. Surface coatings have been used in a variety of forms such as solvent-based waxes, water-based and emulsion-based waxes. The coatings can therefore reduce metabolism and extend the shelf life of the produce by limiting access to oxygen while creating saturation for carbon dioxide.

The latest development to surface coatings like that of edible coatings provide a more consumer-friendly protection. These edible waxes as well as other surface coatings have been made possible by incorporating GRAS (generally recognized as safe) compounds such as food grade sodium bicarbonate or bio-control agents (Kanlayanarat et al., 2009). However, much to the preference of the consumers for additives-free diet, the usage of surface coatings in fresh food applications has been appreciated less. In contrast, for developing countries where refrigeration is not affordable, the use of these coatings is gaining popularity, especially the ones that are of natural origin and for their cost-effectiveness.

9.7.5 REFRIGERATED TRANSPORT SYSTEM

For large-scale commercial operations, refrigerated storage may be used in a cold-chain operation to carry regular consignments from production areas to urban markets and retailers. This can be a highly complex operation requiring expert organization and management (FAO, 1989). A state of the art system of cold storage chain is never complete without an efficient transportation support. All the protocols care, and precautions followed while handling the produce at various stages of cooling and storage will be laid to waste without this unit operation.

The risk of deterioration to occur while in transit magnifies due to several factors and the produce can be spoiled by a number of agents in different ways. No matter the mode of transport involved, the produce must be handled observing three critical points:

- The produce is kept as cool as optimally possible;
- The produce is kept dry, avoiding water retention, moisture build-up for moisture sensitive items;
- Once the produce reaches its destination, it should reach the market as soon as possible, and arrangement should be made such that the time lag is kept to the minimum.

9.7.5.1 TRANSPORT TRUCKS

Transport by land on roadways stays the more preferred and affordable means to shift produces from the warehouses to markets or distribution centers, and it is specially the only means when the site remains largely inaccessible by other means. The vehicles commonly used are open pick-up trucks or larger trucks, either opened or closed. The following care and precautions are to be taken up while transporting via trucks:

- Closed vehicles are to be used only for short transits, for local deliveries from farmers or wholesalers to nearby retailers.
- Open-sided or half boarded trucks should be fitted with canvas to allow easy loading and off-loading at any point around the vehicle by rolling up or moving aside the canvas. Wire meshes can be used in the sides and rear of the truck where pilfering is a problem.

- A secondary white-painted roof can be fixed 8 or 10 cm above the main roof to shield the produce from radiations and keep the produce from heating up (FAO, 1989).
- For long transits, a more elaborate airflow through the produce should be arranged by fitting air intakes in conjunction with louvers.
- For refrigerated truckloads, airtight and moisture-proof containers with power supply and controls can be used, but remains uneconomical for small-scale operations due to high costs.

9.7.5.2 HANDLING AND STORAGE PRACTICES

The shape, condition, and hygienic maintenance of the trucks serve to be important factors in produce transportation. However, mishandling, and improper loading and stowing practices may account for damages and losses:

- A stable and well-ventilated load must be aimed for: a maximum capacity load carried under satisfactory technical conditions (FAO, 1989).
- The size, design, and strength of packages should be suited to the levels of ventilation in the transport vehicle and vibrations and stresses to be acquired in transit.
- Proper supervision while loading and unloading of vehicles should be maintained to check mishandling and ensure the use of automatic and mechanical aids like trolleys, roller conveyors, pallets or forklift trucks to reduce individual handling of packages.
- Stowing should be done such that the stowed packages stay in place during transport and within maximum stowing limits without collapsing or damaging the bottom layers due to the weight of those above.
- Packages should be protected at all times from the elements such as sun and rain.
- Packages should be loaded on dunnage (pieces of lumber or slatted racks) on the beds of vehicles, or on pallets in order to allow the circulation of air around stacks during transport (FAO, 1989).
- If the loads are to be distributed to several locations, the packages should be loaded in reverse order on a 'last one, first off' basis and such that the loads be evenly distributed on the vehicle (FAO, 1989).

As far as driving the loaded vehicles are concerned, rash driving or speeding to make more money can result in produce damage. Only experienced and seasoned drivers should be allowed to take the wheel.

9.7.5.3 RAIL TRANSPORT

Transport by rail is faster and has obvious advantages than road transport. Because of a relatively smoother track in railways than the less dependable roads, the damages incurred during an otherwise transportation over rough roads are within acceptable limits. It is a cheaper mode of transport than road but requiring extra handling while transferring from the source to the train station and later from the station to the destination.

9.7.5.4 WATER TRANSPORT

It is a promising mode of transport for regions inaccessible by the rail or the road. Transportation is rough owing to local inadequate packaging materials and handling.

1. **Inland:** In some countries, this mode of transport is adopted as a means to move produces packed and shipped locally in crates or sacks with no special measures or handling for fresh produce. Very often the vessel is a mixed passenger-cargo craft.
2. **Sea:** For countries of the island community (e.g., the Philippines), short-distance transport of fresh produce in small ships without refrigeration is common. The vessels are passenger cum cargo crafts and the losses are high due to mishandling by porters, inadequate packaging, heating due to unventilated holds or holds near engine room.

A model for organized and efficient mode of sea transport is the refrigerated transport of loads such as bananas, although further improvements can be made by modest investment by the small-scale enterprises (FAO, 1989).

9.7.5.5 AIR FREIGHT

As with shipping, the international trade in the airfreighting of high-value exotic crops is generally well organized. In some countries where road links are poor (e.g., Papua New Guinea), produce is carried by air from production areas to urban markets (FAO, 1989). The airfreight transport demands high costs and is subjected to heavy losses due to overall glitches in packaging and handling (poor packaging, exposure to elements, temperature fluctuations, etc.). Moreover, passengers are preferred over consignments and the flight delays due to bad weather are unpredictable. Road transport is a better option in this regard.

9.8 CONCLUSION

The basic aim and principle of a cold storage facility is to keep a check on the metabolic processes or the spoilage factors such as ripening, maturation, decay, and final deterioration in working during storage, transport, and distribution so that these factors are maintained within acceptable limits. The ultimate benefit goes to bridging the gap between the unwavering increased demand due to an exploding global population and the variable supply of produce all the year-round.

This requirement to curve the deficit in supply can only be materialized through properly organized and designed commercial cold storage systems as a major portion of produces go wasted largely owing to absence or inefficient produce handling, storage, and transportation systems with the necessary infrastructure cum machineries.

The success of a low temperature storage system depends on how well the produce was pre-cooled prior to refrigeration. This involves removing most of the field heat as much as possible and in a little time of processing. Pre-cooling is a step of preparing the harvested produce in a way to reduce the load of heat removal from falling entirely on refrigeration systems alone. Traditional practices along with conventional methods are used depending on produce sensitivity for temperature and humidity.

Temperature and humidity management along with produce and hygiene management by the food handlers are critical to delivering the effectiveness of cold storage chain. It can be brought to attention that those responsible for handling food after harvest right up to the point of delivery holds the key to the execution of this system. The food handler should be equipped with basic training, general know-how, and technical knowledge of the cold chain.

Supplementation systems such as controlled and modified atmosphere storage systems that function in sync with cold storage are preferred where sustainability is not a question. Low cost evaporative systems and surface coatings are further additions gaining popularity for their cost-effectiveness. Storage systems demand to be well organized and prepared for mixed commodity storage, if the need ever arises. Disorders are to be kept at bay through the entire operation by managing temperature fluctuations or preventing it entirely.

A cold chain is never complete without refrigeration equipped or hygienic unrefrigerated transport systems either by road, rail, water, airfreight or variable transport including a combination of these. The mode of transportation may depend on the suitability, the availability, the best route, and the location accessibility. Precautionary measures and hygienic practices accompany such transport systems.

The need to establish commercial cold chain facilities to meet the increasing volumes of production of fresh produces is beyond doubt a matter of global necessity. So, prior to planning and establishing such systems, the factors that contribute to the working efficiency of these systems need to be carefully evaluated and considered for their applicability case by case, based on their scales of operation, the capacities involved and the resources available for utilization.

KEYWORDS

- **carbon dioxide**
- **controlled atmosphere**
- **first in first out**
- **generally recognized as safe**
- **hydrochlorofluorocarbon**
- **liquid nitrogen**
- **modified atmosphere**

REFERENCES

Barbosa-Cánovas, G. V., Altunakar, B., & Mejía-Lorío, D. J., (2005). *Freezing of Fruits and Vegetables: An Agribusiness Alternative for Rural and Semi-Rural Areas* (Vol. 158). Food & Agriculture Org.

FAO (Food and Agriculture Organization of the United Nations), (1989). *Prevention of Post-Harvest Food Losses Fruits, Vegetables, and Root Crops a Training Manual*. Pre-harvest factors in produce marketing: Training series 17/2.

Gustavsson, J., Cederberg, C., Sonesson, U., Van, O. R., & Meybeck, A., (2011). *Global Food Losses and Food Waste* (pp. 1–38). Rome: FAO.

Kanlayanarat, S., Rolle, R., & Acedo, Jr. A., (2009). *Horticultural Chain Management for Countries of Asia and the Pacific Region: A Training Package*. FAO.

Kitinoja, L., & Kader, A. A., (1995). *Small-Scale Postharvest Handling Practices: A Manual for Horticultural Crops* (3rd edn.). University of California-Davis, California.

LalBasediya, B. A., Samuel, D. V. K., & Beera, V., (2013). Evaporative cooling system for storage of fruits and vegetables: A review. *Journal of Food Science and Technology*, *50*(3), 429–442.

McGregor, B. M., (1989). *Tropical products transport handbook*. Agriculture Handbook No. 668. Revised Edition. Office of Transportation, US Department of Agriculture, Washington D.C.

Mitchell, F. G., Guillou, R., & Parsons, R. A., (1972). Commercial cooling of fruits and vegetables. *Commercial Cooling of Fruits and Vegetables*, 43.

Persson, P. O., & Londahl, G., (1993). Freezing technology. *Frozen Food Technology*, 20–58.

Salin, V., (2014). *2014 IARW Global Cold Storage Capacity Report*. International Association of Refrigerated Warehouses, Alexandria, VA.

Shewfelt, R. L., (1990). Quality of fruits and vegetables. *Food Technology (USA)*.

Stoecker, W. F., & Jones, J. W., (1982). Compressors. *Refrigeration and Air Conditioning* (2nd edn.). McGraw-Hill, Inc.: New York, NY, USA.

Fresh Plaza, (2014). *Global Production of Fruits and Vegetables Increases by 9.4%*. https://www.freshplaza.com/article/2117700/worldwide-production-of-fruits-and-vegetables-increased-by-9–4/ (accessed on 12 January 2021)

CHAPTER 10

SANITATION AND HYGIENE PROCESS IN STORAGE

MEENAKSHI GARG,[1] SADHANA SHARMA,[2] RAJNI CHOPRA,[3] and SUSMITA DEY SADHU[1]

[1]*Bhaskaracharya College of Applied Sciences, Delhi, India, E-mail: meenakshi.garg@bcas.du.ac.in (M. Garg)*

[2]*National Institute of Food Technology and Entrepreneurship Management, Haryana, India*

[3]*Institute of Home Economics, New Delhi, India*

ABSTRACT

Warehouse storages are budding ground for activities hampering quality and safety of food produce. No matter, if the food has been produced or manufactured under good hygienic conditions, if the environment surrounding the food during storage is not neat and clean, hazards will come in. Therefore, sanitation, and hygiene of the premises where the food is stored is necessary and should not be overlooked. This chapter sheds light on the understanding of sanitation and hygiene, why it is important with respect to food warehouses, and how it can be well achieved. It disseminates information helpful to the manufacturers, distributors, and the retailers who are willing to ensure the wholesomeness of their produce during storage. This will assist in cutting down on food losses occurring during poor storage conditions, thus indirectly fetching economic benefits and food security for the country.

10.1 INTRODUCTION

Food concerns all of us. Safe food is very important because in the end, we all are consumers. For ensuring food safety, the sources of food contamination need to be determined at every step of the food supply chain from conception to consumption. One of the areas of food safety is hygiene and sanitation. No matter, if the food has been produced or manufactured under good hygienic conditions, if the environment surrounding the food during storage is not neat and clean, hazards will come in. Therefore, sanitation, and hygiene of the premises where the food is stored is necessary and should not be overlooked. This chapter sheds light on the understanding of sanitation and hygiene, why it is important with respect to food warehouses, and how it can be well achieved. It disseminates information helpful to the manufacturers, distributors, and the retailers who are willing to ensure the wholesomeness of their produce during storage. This will assist in cutting down on food losses occurring during poor storage conditions, thus indirectly fetching economic benefits and food security for the country.

Storage is a major place for the effective marketing function, which involves both preservation and storage of goods for a long time, i.e., from production to consumption (Storage temperatures for fruits and vegetables, 2015). Good food storage not only ensures freshness, but also reduces waste. Food storage allows food to be eaten immediately and for some weeks to months also after harvest. The stored foods can be flowed continuously in the market from production to the consumption time period directly. Fruits, vegetables, and their products which are stored must be safe from drastic weather conditions and from foreign contaminants which could lead to food wastage and spoilage. Stored products should be protected from polluted environment and pests. Storage godowns and warehouse facilities should be electrified, well lighted, and have a system of fire prevention according to rules and regulations. When a warehouse receives a food delivery, the lot must be stored in the appropriate storage area to prevent spoilage and wastage. The storage instructions should be informed properly to the storage personnel to facilitate good storage practice. Depending on the type and quality of food being stored, several kinds of storage options are available. Traditional and domestic skill in the form of food logistics, are important for industrial and commercial activity. The major problem of food security is bad food storage practices. For food security, preservation of food, food storage, effective transport, and timely delivery of horticulture produce to consumers are important, and the main purposes of this are as follows:

- For distribution of products to consumers, storage of harvested and processed plant food products should be available.
- Storage is important for the production of fresh produce on a continuous basis, which enables a better balanced diet throughout the year.

- One should be prepared for catastrophes, emergencies, and periods of famine and food scarcity.
- Quality of perishable, semi-perishable, and nonperishable products can be protected from deterioration by storage.
- Stabilization of prices helps to adjust demand and supply of food products.
- For performance of other marketing functions, storage is necessary for some period.
- Protection from animals or theft is also important.

10.2 TYPES OF STORAGE

Storage of food can be classified on the following basis such as: (i) type of ground structures like surface ground or underground; (ii) level of storage according to volume, e.g., small scale storage or large scale storage structures; (iii) type and quantity of food to be stored viz., Dry storage at room temperature, Cold or refrigerated storage, and frozen storage (Storage temperatures for fruits and vegetables, 2015). These are briefly explained below:

1. **Underground Storage Structures:** These are dugout structures, i.e., sand is removed from the ground to a certain depth similar to wells and sides were plastered with cow dung and Lined with stones, sand, and cement or polyethene (Figure 10.1).

   Shape of these structures can be circular or rectangular. The storage capacity of the structure depends upon the size. These storage structures provide safety from various external sources such as theft, rain, or wind and can be easily filled up due to the effect of gravity.
2. **Surface Storage Structures:** Food grains can be stored on ground surface in bag storage or in bulk storage.

   i. **Bag Storage:**
      a. Bags bear a definite quantity and are easier to load or unload. These can be sold or dispatched without difficulty.
      b. Bags are separated on the basis of identification marks labeled on the bags.
      c. If in any bag there are chances of infestation, it can be removed and treated easily.
      d. The surface of the bag is exposed to the atmosphere; hence chances of sweating at bulk or loose storage is minimum (Figure 10.2).

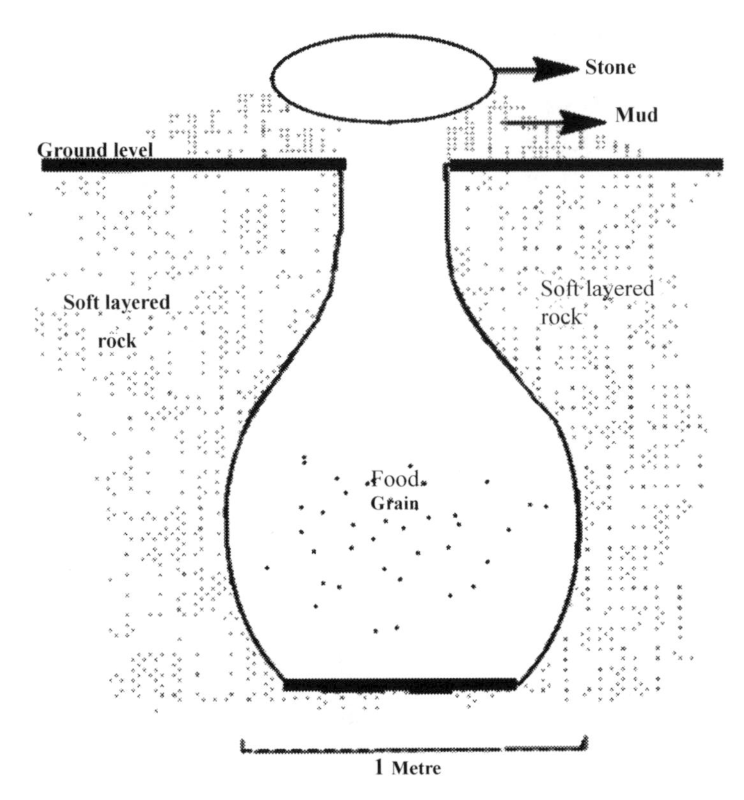

FIGURE 10.1 Underground storage structures.

Source: Proctor Hall et al. (Anonymous, 2015).

FIGURE 10.2 Surface storage structure.

e. Exposure of the grains to external threat is reduced because exposed peripheral surface area/unit weight of grain is very less. In addition, conditions in the deeper layers of the grain are airtight; hence there is the lesser incidence of any pest infestation.

Above mentioned are the conventional storage techniques, under Government of India initiatives, various other advanced storage facilities have been developed at farm level. These include structures for small scale and large-scale storage:

1. **Small Scale Storage:** PAU bin, Pusa bin, Hapur Tekka.
   - **PAU Bin:** It is a galvanized metal iron structure designed by Punjab Agricultural University (PAU) Ludhiana, having storage capacity is 1.5 to 15 quintals.
   - **Pusa Bin:** It is constructed with bricks and mud with a polythene film lining embedded within the bin walls (Figure 10.3) (Anonymous, 2015).

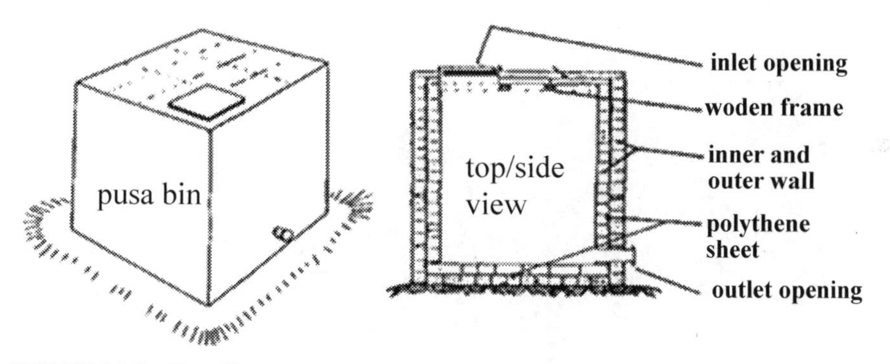

FIGURE 10.3 Pusa bin.

   - **Hapur Tekka:** It is cylindrical in shape and its cloth is made from rubber material supported by bamboo poles on a base which is a metal tube. Tube has a small hole and perforated in the bottom (Figure 10.4).
2. **Large Scale Storage:** Cover and Plinth (CAP) and Silos:
   - **CAP Storage (Cover and Plinth Storage):** In this type of storage, from ground to a height of 14" brick pillars are constructed with grooves. Wooden crates are fixed in grooves for the stacking of

bags of food produce. It can be constructed in the limited time period of few weeks only. It is a cheaper method of food storage on a large scale (Figure 10.5).

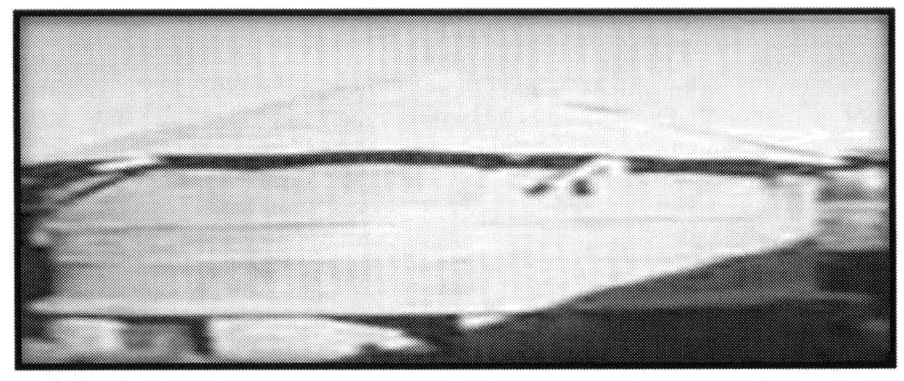

FIGURE 10.4 Hapur Tekka.

FIGURE 10.5 CAP storage.

- **Silos:** In these types of structure (Figure 10.6), the fruit and vegetable produce are first unloaded on the conveyor belts in bulk and then by mechanical operations transferred to the storage structure. For single silos, the capacity of storage is about 25,000 tons (Anonymous, 2015).

FIGURE 10.6 Silos storage with conveyor belts.

10.3 CRITERIA FOR GOOD STORAGE

Proper storage environment must be maintained to ensure the safety and wholesome of produce during storage; only a good storage structure cannot provide quality and safety. For a storage operation, three important aspects must be considered, i.e.: (i) temperature of storage; (ii) humidity of storage; and (iii) presence of an infestation.

10.3.1 TEMPERATURE PREVAILING IN STORAGE PREMISES

Temperature of food storage will depend on the nature, time period, and type of food to be stored. Generally, bacteria are inactivated at freezing point, i.e., at 0°C or below. They start becoming active in between 1.7°C and 4.4°C but grow slowly. With the increase in the temperature, at 37°C which is the normal body temperature, the bacterial activity increases. Recommended storage temperatures and storage life in weeks for different food materials are shown in Table 10.1.

Most perishables products should be kept at temperature below 10°C and semi or non-perishable foods should be stored in between 10°C and 20°C to resist microbial growth. Table 10.2 represents the different storage types with examples of the suitable produce. The elimination of pathogens (disease-producing organisms) can be achieved by maintaining the correct product temperatures of high-risk foods. For this knowledge of the danger zone and the temperature range in which the multiplication of pathogenic bacteria is must.

TABLE 10.1 Ideal Food Storage Temperatures for Different Horticultural Produce

Fruits/Vegetables	Storage Life in Weeks	Ideal temperature in °C
Apple	12–52	–1 to 2
Tomato	3	13
coconut	8	0
Lemon	26	10
Banana	4	13
Pumpkin	18	13
Mango	4	13

Dry Food Storage	Cold Storage	Freezer Food Storage
Dry food storage is for those foods which do not support bacterial growth in their normal conditions. Such foods have a long shelf life and can be kept at room temperature. E.g., coconuts, ginger roots, sweet potatoes, watermelon, dried beans, and peas.	Cold storage is for foods that must be kept in the refrigerator or cool room below 5°C. E.g., grapes, bell pepper, citrus fruits.	Frozen storage is for foods that need to be kept at minus temperature. E.g., pears, apples.

10.3.2 HUMIDITY IN THE FOOD STORAGE ENVIRONMENT

The rate of growth of microorganisms depends upon the moisture content of the air present in the storage area. Foods that have low moisture content can be safely stored at higher temperatures. Such foods when exposed to the atmosphere they have the ability to absorb moisture. Generally, in humid storage conditions, microbial growth and insect infestation is more common. Foods which absorb moisture, becomes perishable easily, and they require a lower temperature for food storage. Humidity deteriorates the quality of fresh fruits and vegetables (F&V) and make them unfit for human consumption. Due to unsafe consumption, substantial cases of diarrhea have been noticed, because fresh produce is directly contaminated with microorganisms. In tropical countries like India, Pakistan, etc., where humidity in the environment changes drastically, the best method to protect low moisture horticultural produce is to store the stock in air-tight containers and in small quantities so that it can be consumed fast and quickly enough to prevent deterioration.

10.3.3 PRESENCE OF INFESTATION

The infestation within or near the food storage premise may contaminate the food, and food may become unfit for humans to consume, even if sanitation and hygiene is checked and monitored carefully. Rodents, insects, flies, and cockroaches are commonly present in storehouses. Food Poisoning may occur when these pests deteriorate fruit and vegetables and humans consume these infected produce. Contamination can also occur due to pests droppings they leave behind. Flies habitually while feeding sit on food, through their saliva or dirt or foreign material present on the fine hairs on their body, might be discharged in the food. Sometimes they put eggs of parasitic worms on fresh produce. Pests enter in food through cuts, defective drains, holes, or bags or from waste material stored outside or within storage premises. Some pests like cockroaches remain in unclean, dark corners, crevices of walls, under equipment, in drawers, behind cupboards, etc. Due to this reason, it is difficult to find any cockroaches in the daytime. A proper inspection is therefore important at night.

10.4 SANITATION AND HYGIENE

Sanitation refers to creating and maintaining hygienic and healthy conditions (Food Storage Warehouses, 2017). It is a dynamic and continuous operation which needs to be dealt with every now and then. Food sanitation may be defined as the application of science in providing wholesome food that is handled or cared for and supervised in a hygienic and safe environment by healthy and skilled food handlers to prevent contamination from food pathogenic and spoilage microorganisms. It is, therefore, an applied sanitary science related to processing, preparation, and handling of food (Food Storage Warehouses, 2017). It is nearly impossible to consider food sanitation and sanitation of the environment (in which food products are manufactured, stored, and prepared) as two different aspects (Food Storage Warehouses, 2017). Sanitizing is the process used to remove or minimize the number of microorganisms flourishing on the surface. Sanitizing cannot be accomplished without cleaning and a good pest control program (Anonymous, 2015). Cleaning refers to the removal of soil, food residue, dirt, grease, or other objectionable matter (Food Storage Warehouses, 2017). Standard practices of high-level sanitation and hygiene must be implemented in every aspect of the food processing and manufacturing (Anonymous, 2015). Sanitation and hygiene includes all aspects like personnel, premises, utensils, and storage area that can spoil food product (Anonymous, 2015). Employees shall be trained, well instructed and motivated to make reporting on a regular basis to their immediate supervisor regarding conditions related to personnel, plant, or equipment that they think may adversely affect the product quantity and quality (Anonymous, 2015).

Hygiene refers to the conditions and practices that assist in maintaining health and preventing the spread of disease (Anonymous, 2015). Good Hygienic conditions and measures are important to ensure the safety and suitability of horticulture produce at all stages and levels of the food chain (Food Storage Warehouses, 2017). This covers manufacturing process, personnel involved, premises, equipment, and handling of materials from starting raw materials to the final or finished products (Anonymous, 2015).

Thus for an effective care and handling of food products, a high level of sanitation and hygiene is mandatory both on the part of the manufacturing facility and on the employees who work there (Anonymous, 2016). A successful running food-grade warehouse must have a superior quality sanitation schedule and process of cleaning/sanitation should be mentioned in detail and must be followed strictly (Anonymous, 2016).

10.5 IMPORTANCE OF SANITATION AND HYGIENE IN STORAGE

In order to maintain product acceptability and compliance with regulatory requirements, effective and rigid sanitation practices are essential in storage facilities. The necessity of warehouse sanitation and cleanliness cannot be ignored by companies that demand the best, clean, and modern warehousing and storage premises that are free of foreign debris that can contaminate stock lines (Food Storage Warehouses, 2017). In many of Asian countries, food storage warehouses are common for fresh, canned, and frozen F&V due to the tropical climatic conditions. Keeping a large warehousing facility, one should take care of proper equipment, organization, and supplies (Food Storage Warehouses, 2017). The most important is warehouse flooring, always keep it clean from debris of packaging materials, fruit, and vegetable waste and accumulated dust is an ongoing and necessary task. Even if stocks are stored on the premises for a long time, the area should be kept as clean as the rest of facility, e.g., between pallets, container bays, repacking areas, offices, restrooms, and walkways, all are all thoroughly maintained and cleaned (Food Storage Warehouses, 2017). Horticultural produce should be stored on clean racks/ pallets that are reasonably well above the floor level and are away from the wall to facilitate cleaning and which prevent harboring of insects, pests, and rodents. Raw materials should be sampled properly during the receiving operation to verify and check that they are not infested with any type of organism like insects, molds, rodents, or other unacceptable contaminants and proper housekeeping practices should be followed. Storage areas require routine inspection to check microbial and pest infestation. Inspection and cleaning frequency of depends upon the temperature and humidity of storage. Premise should be neat, dry, and clean and free of any puddle water. To facilitate good sanitation, premises used for the storage of F&V shall be of suitable design and construction (Anonymous, 2015).

Maintenance of sanitation by sweeping and cleaning aisles and stock locations is well understood. However, challenges with sanitation are not limited to cleaning pallet debris and sweeping empty pallet locations. Seepage from raw product, allergen spills, and fluids from stored food and other products can lead to risks to stored products and environmental contamination issues. To eliminate the potential for cross-contamination, implementation of proper sanitation and hygiene techniques are important (Anonymous, 2015). Since the enactment of the FDA's Food Safety Modernization Act in 2011, food storage warehouses have been under scrutiny with increasing expectations and regulation and in regard to holding, handling, transferring, and storing safe, wholesome food. Food storage warehouses have a set of guidelines and food safety requirements that can be daunting without proper education, support, and resources. In order for the food industry to offer a true guarantee of wholesome and safe food, emphasis, and support should be provided to the distribution centers who act as one of the important and last stops on the farm-to-fork chain (Anonymous, 2015).

10.6 WAREHOUSE DESIGN

A "Food Storage Warehouse" is any area, establishment, building, room, facility, or place, in part or whole, where food is kept, stored, or held for wholesale distribution to other wholesalers or to retailers, restaurants, or any other distributor or to the ultimate consumer. "Food storage warehouse" does not include elevators or fruit and vegetable storage and packing houses that pack, store, and ship fresh fruit and vegetables. Food warehouses are a critical and important part of the food distribution system. Warehouses can be of many kinds:

1. Those that store food items either refrigerated or canned.
2. Those that store non-food (FMCG products, furniture, etc.), items, and chemicals (Anonymous, 2015).

Food grade warehouses serve as a link among the fields of the farmer, the processor, the retailer, and the consumer (Anonymous, 2015). Storage facilities should undergo rigorous evaluation and must adhere to the guidelines to continue operation. Food grade warehouses include dry storage, cold or frozen storage, and chilled or refrigerated storage depending upon the type of horticultural commodity to be stored (Anonymous, 2015). These food-grade warehouses maintain proper health and sanitation in order to protect the food within (Anonymous, 2015). A warehouse building should not have:

- Leaks in the edges of the foundation, roof or walls.
- Weeds, trash, rodent tracks/burrows or standing water, around the perimeter of the building.
- Cleaning agents like detergents, pesticides, fungicides, insecticides, and other chemicals in the area where F&V are stored.
- Any types of cracks or holes in the windows or the window frames.
- Damage to the exterior and interior of the building like cracks, holes, open pipes, etc.

Warehouse location also plays a significant role in facilitating sanitation and hygiene during storage.

A food-grade warehouse should:

- Avoid any type of environment pollution and activities that can produce any objectionable odor, dust, smoke, fumes, pollutants, and chemical emissions.
- Avoid spaces which are prone to pest infestation and areas where waste is dumped or cannot be removed properly and efficiently.
- Be away from direct access to any residential area.
- Provide adequate lighting and ventilation in the premises and walls should be whitewashed or painted properly.
- On floor drainage system should be adequate, and it should be easy to clean, wash, and disinfect.

- Screens with wire mesh on all doors and windows and fit doors having automatic closing springs should be present to prevent entry of any insect. The screens, mesh, and frame should be easy to clean and wash.

In the warehouse, products that can contaminate other products due to their strong odor should be stored separately. Usually the food packaging material is sensitive to odors and should be used judicially to avoid absorption of off odor. For further safeguarding horticulture produce against contamination, food-grade warehouse facilities must follow McKenna's 4 Principles for Food Grade Storage (Anonymous, 2015). Personal Hygiene and Training, Pest Control, Sanitation schedule, and Lot traceability are the main four principles.

10.6.1 PERSONAL HYGIENE AND TRAINING PRINCIPLE

This principle states that food-grade storage warehouses employees must wash their hands with company-supplied effective soaps in sinks or washbasins having good hygienic drying systems. Record of employees related to quality awareness training (hygiene, crisis management, food safety) must be maintained and updated (Anonymous, 2015). Personal hygiene also refers to the cleanliness of a person's whole body against microorganisms which causes disease and from dirt, filth, and other foreign material.

All personnel, prior to employment shall undergo health examinations (Anonymous, 2015). During their employment period, they should also undergo health checkups periodically, which must include examinations related to the tasks that they have to perform in the storage house. For example, general health examination for all personnel (Anonymous, 2015). All personnel shall practice good personal hygiene. They shall be trained in the practices of personal hygiene. Appropriate practices in personal hygiene play a vital role in sanitation during the storage process. Related sanitation and hygiene procedure and instructions should be written and displayed in the entry of the warehouse (Anonymous, 2015) such as follows:

- Bath/Shower-Wash off dead skin cells and disease-causing microorganisms.
- Brush teeth-Brushing helps to clear up the residues of food, which is a major source of microorganisms.
- In the storage premises, dirty clothes, footwear and gloves can carry microorganisms. Hence complete uniform including gloves and shoes should be clean. Polish and artificial nails can fall hence fingernails

should be kept clean, trimmed, and nail polish should be removed. Person suffering from any apparent illness or open lesions shall not be allowed to handle any food item during storage it may affect the quality of products (Anonymous, 2015).

In addition to personal hygiene, training of employees is an essential component in accomplishing warehouse food safety. All employees should be aware of all the hazards of compromised or adulterated foods and should attend employee-training programs from time to time. Supervisors and managers should be given trainings to understand the role of leadership which they have to play in sanitation and should be updated regarding changes in industry regulations. All food handlers are trained in good manufacturing practices (GMPs), standard sanitation operating procedures (SSOP'S), and to 'clean as you go.' As a food business, sanitation, and warehouse food safety is a way of life.

10.6.2 PEST CONTROL

Pests such as rodents, insects, etc., are present in godowns, these pests must be controlled or killed before the sanitation step. Hence pest control is the reduction in number or elimination of mice and rats, flies, cockroaches as well as weevils and other insects that can infest food products (Schuler et al., 1999). If all the pests are eliminated, cleaning followed by sanitizing is sufficient. This is because the pests will re-contaminate any surface that may have been sanitized (Schuler et al., 1999). There are three measures of pest controls. The first and most important is the exclusion of pests from our facility (Environmental Control). This is accomplished by making building pests proof. The second measure is to use trapping and poisoning (Physical Control) for exterior perimeter. The third measure is interior interception. Eighty percent of Warehouse Food Safety is good housekeeping. Substances that ensure the prevention or elimination of pests like rodents, birds, insects, ants, and other animals must be placed or kept around the vicinity of the building (Anonymous, 2015). Updates to the pest control routine and any changes must be noted quarterly (Anonymous, 2015). An effective sanitation program is the foundation of any pest-management program. Many of the pests issues encountered during food storage originate from outside the manufacturing plant and storage premises. Pests enter through the loopholes in the storage structures, causing losses and contamination problems in warehouses. Thus for an efficient pest control program, it is important to understand the way how pests invades the storage area, their biology, and harborage behavior. Residual food and debris attracts insects, birds, or rodents.

1 **Sanitation Schedule:** The warehouse facility should be cleaned and tidy at all times (Anonymous, 2015). Food storage must require the highest standards of sanitation procedure to prevent spoilage and avoid product waste (Anonymous, 2017). Misfortunes happen, and the schedules get overloaded, employees get overtaxed and have to cut corners to meet deadlines, and these things can start to impact the sanitation of your products. It should be ensured that schedules are kept to and that every possible step is being taken to clean your storage facilities and prevent external contamination of food products (Anonymous, 2017).
2. **Lot Traceability:** The warehouse must have a proper system for tracing date codes and a lot of products. To ensure this, a "first-in, first-out" method of inventory rotation can be followed. These four principles and the strict quality control standards ensure that the food items within will never be jeopardized (Anonymous, 2015).

10.7 INTEGRATED PEST MANAGEMENT (IPM) PROGRAMS IN FOOD WAREHOUSE

Integrated pest management (IPM) program is a pest control strategy that combines physical, chemical, and biological methods to achieve comprehensive pest control (Heaps, 2006). It is a long-term solution where the customer and pest control experts work together to effectively prevent and control pests. Insects such as biscuit beetle, rice weevil, and rust-red flour beetle, particular stored product insects might be imported accidently with food commodities, where as wood-boring insects might be hidden in wooden boxes or furniture. The commercial food products stored in the warehouses before and after manufacturing prior to distribution of final product need attention for the sanitation and pest control. It is important for the manager of warehouse or business owner to understand the importance of pest control. Staff of warehouse must have the knowledge whom to contact at the time emergency, and that the pest control contractor is fully up-to-speed with the latest legislation, control methods and technologies. When one has to implement a successful IPM solution, the construction, age, layout, size, condition, access, location, and surrounding grounds, as well as the warehouse design itself, must all be taken into considerations for pest control (Anonymous, 2015). This IPM involves multiple key components including sanitation, insects trap and monitoring building design and pest proofing, storage practices, pesticide application, and fumigation.

10.7.1 SANITATION

Effective sanitation program outside and inside the warehouse is the first step in managing stored product pests (Nijasure, 2013). Areas which are outside the warehouse must be kept neat and clean to prevent entry of insects or rodents. Clutter or waste material as if unused pallets, used packaging material, used gunny bags should be removed thoroughly (Nijasure, 2013). The warehouse manager must check signs of an infestation in goods coming into the building. Shredded material, waste packaging material, rotten food products, piles of dust after removing goods from the pallets are the major signs of unwanted pests in the goods that are about to be stored. Supervisors should also observe unusual smells, signs of droppings, smear marks on the walls, and gnawing on any woodwork. Any leftover packaging material and the empty space must be thoroughly cleaned. Insects and rodents entry can be prevented by creating a vegetation free zone or barrier around the outside area of the warehouse. Daily and Periodical (weekly or longer) are the two types of cleaning duties. Master cleaning schedule for sanitation should be planned (Nijasure, 2013). Cleaning of washrooms and another facility should be done to avoid contamination of food products (Anonymous, 2015). Hazardous material, detergents, sanitizers, and other supplies should be handled carefully. Cleaning compounds and hazardous materials should be stored separately from food products and kept in original containers. Infested material must be disposed of in an adequate manner. Toilets and dressing rooms should be clean, ventilated, and separated from the storage area. Hand-washing facilities like soap, hot water, and sanitary towels should be there (Anonymous, 2015).

10.7.2 BUILDING DESIGN AND PEST PROOFING

Pest proofing provides a solution for pest control and reduces the requirement for repeated chemical control. From surroundings pests may intrigue the storage premises (Nijasure, 2013):

- **Exterior Area:** Most of the pest problems generate outside the food plant and can be managed by removing conditions which promote pests harborage around the structure. Exterior pest-proofing techniques involves:
    - **Entrance and Exit Doors:** To prevent entry or movement of flying pest's air curtains or plastic stripes should be installed. Screen doors should be fitted. Automatic door closing mechanism must be used. Metal doors are preferred as wood is susceptible to weathering and gnawing by rodents.
    - **Windows:** Should be fitted with a mesh and properly designed.
    - **Foundation:** All openings which have an area greater than ¼ inch should be sealed to prevent rodents.
    - **Drains and Vents:** These are entry points for insects and rodents and hence must be screened and cleaned adequately.
    - **Utility Lines:** Plumbing pipes, electrical conduits, etc., all are convenient runways for entry of pest into the building. Gaps must be sealed properly.
    - **Roofs:** All openings in the roof must be properly sealed and tightly fitted to prevent any entry of pests.
    - **Shipping and Receiving Docks:** Overhang should be constructed for exterior docks, this acts as a barrier for rodents.

Area inside the premises, specially building design and pest-proofing of whole food warehouse reduces chances of proliferation of pests that have gained access into the building. Ceiling, wall, and floor should be free of cracks.

10.7.3 STORAGE PRACTICES

Inspection aisles next to the wall should be painted white, and products are to be stored away from walls (Nijasure, 2013). Proper management of storage pallets used inside food warehouses is important as they harbor various pests. Infested packaging materials should be immediately discarded when noticed as it acts as breeding grounds for insects. Proper cleaning, quarantine, and pest management procedures must be implemented to prevent infestation. Good warehousing practices (GWP) like first in first out (FIFO) principle, etc., should always be implemented and followed (Nijasure, 2013):

- **Insect Traps and Monitoring (Nijasure, 2013):** Locating the source of infestation and identifying the insects are important in view of a safe and sound storage process. Sensitive monitoring tools for various pests, like species-specific Pheromone traps are used to determine the location and intensity of insect infestations.
- **Pesticides Application (Nijasure, 2013):** Pesticides application includes space treatment, fumigation, and the application of residual sprays. The list of pesticides approved by the Central Insecticide Board must be refereed while using pesticides in the food warehouse, and for correct use of pesticides, information mentioned on labels must be referred. Applications of pesticides are broadly classified as residual and non-residual insecticides. Pesticides are sprayed always in non-food areas The places where insect pests have been observed, spot pesticides are applied there using fogging or ultra-low volume misting. Application of pesticides should be done by using appropriate personnel protective equipment and by licensed and trained pest control operators (PCOs). Care should be taken that pesticides should not contaminate the F&V stored in the warehouse.
- **Fumigation (Nijasure, 2013):** All fumigants are generally gases which are quickly disperse to reach the target organism. Fumigants used are insecticide and can kill insects in enclosed area. If used properly, no residues of insects are left behind. Methyl bromide and phosphine fumigants are registered in India. Both methyl bromide and phosphine are restricted pesticides because of extremely toxic nature. Methyl bromide's use is legally restricted only for quarantine and pre-shipment fumigation in India, is an ozone-depleting gas, governed by the Directorate of Plant Protection, Quarantine, and Storage which sanction and issue licenses to fumigators.

Selection of a pest management company carefully is very important for exporters/warehouse owners and the company must have all the valid government licenses and approvals. The company should have well qualified, experienced, and trained personnel. They must have the knowledge of Indian fumigation standards, and quarantine procedures. They should follow the guidelines laid down by the Australian, European, and American plant protection and quarantine authorities to perform these dangerous services. Shortcuts are not advisable as it lead to huge losses for exporters (Nijasure, 2013). Rodenticide, insecticides, fumigating agents, and sanitizing materials when used must not contaminate the raw materials, packaging materials, or finished products stored in the warehouse (Anonymous, 2015).

10.8 EQUIPMENT AND PRODUCTS FOR ENSURING SANITATION

One of the most important areas in the warehouse is the floor- and one of the toughest to clean. With the use of heavy equipment, the floor can get marked or scratched repeatedly. The constant loading, unloading of goods makes it challenging for machine operators to operate around (Anonymous, 2015). Moreover, floor care is a challenge, given the marks by heavy equipment. Some companies feel that warehouse contractors should consult cleaning companies about type of cleaning agents for that particular type of flooring to use to minimize the issues regarding floor maintenance (Anonymous, 2015). The method for cleaning of warehouses is different for machine maintenance as compared to floor cleaning. Nowadays, several companies have started using green chemicals in warehouses where the products are sensitive to toxins. Equipment is becoming battery-operated and environment-friendly. In fact, many of machines have 'Eco! Efficiency' tags (Anonymous, 2015). This means that these appliances reduce the cost, use of chemicals, water consumption and electricity (Anonymous, 2015).

10.9 WASTE MANAGEMENT

Waste material shall not be allowed to accumulate anywhere in the warehouse premises (Anonymous, 2015). It shall be collected in suitable receptacles for removal to collection points outside the warehouse building and disposed off safely in a sanitary manner at regular and frequent intervals. Waste bearing container should be covered (Anonymous, 2015). There should be no harborages and breeding areas for pests at the site of storage. Properly closed areas should be used for the storage of waste materials. Waste storage must be done in a manner that it does not contaminate either the storage area or the environment both inside and outside the warehouse establishment. To avoid pest problems, proper waste disposal mechanism should be followed.

10.10 MONITORING AND INSPECTION OF STORAGE PREMISES

Implementation of regular monitoring and inspection of storage premises is necessary for ensuring proper maintenance of sanitation and hygiene during the storage process. Wherever needful, arrangements for monitoring temperature and relative humidity (RH) of the storage place should be made. FIFO (first-in-first-out), FEFO (first expire first out) stock rotation system for storage of packaged food products should be followed, whenever applicable. Some cleaning procedures may need to be validated to confirm the procedures meet requirements. Others simply require verification that they are conducted in the manner intended in order to meet the goals. A record of the cleaning and sanitation program must be observed and kept securely. A practical analysis of the work environment should involve a variety of warehouse examinations to identify current and potential hazards. It may include a thorough baseline survey to review operational methods and personnel practices and individual potential hazards; management of change; a formal written self-inspection program every two weeks and the daily inspections. Chemicals used for cleaning should be used and handled carefully in accordance with the instructions given by the manufacturer and should be stored away from horticultural produce, in clearly identified containers, to avoid risk of any type of contamination. Arrangements for storage of food and its ingredients should be made properly. These should be segregated adequately and labeled properly. During storage operation systems must maintain time and temperature control. Containers made of non-toxic material should only be used for storage purpose. Cold storage facility should be provided, wherever necessary. Thus an inspection program is necessary for a good preventive sanitation program (Schuler et al., 1999). The industry should be preparing its own respective Inspection forms that fit the needs of storage of their items. A checklist may be designed which may include a variety of items that should be inspected on a routine basis. Following is a checklist:

- **Dry and Packaged Food Storage:** Proper storage practice, pest evidence absent, empty container storage and good housekeeping.
- **Damaged Good Storage:** Segregation and repackaging, proper housekeeping and maintenance of returned goods, adequate handling program.
- **Refrigerated area:** Pest evidence absent, condensation absent, cleaning satisfactory. Moreover, in order to protect foods, written procedures (Schuler, 1999) to ensure safe food storage should cover well information pertaining to:
    - o Adequate storage space.
    - o Storage of chemicals and cleaning items (nonfood items) separately from food.
    - o Securing of poisonous or toxic chemicals which are not held for sale.
    - o Distinction of raw animal foods from horticulture produce.

- o Protection of foods from contamination.
- o Separate storage area provision for salvage items.
- o Procedures for handling salvaged, damaged, expired or contaminated foods (include timely disposal of damaged, spoiled, or expired products). Disposition of these food items must also be documented.
- o Rotating stock as per FIFO principle.
- o Provision of adequate lighting and ventilation.
- o Properly labeling of food.
- o Cleaning and organizing of food storage areas.
- o Maintenance of proper food storage temperatures.
- o Maintenance of an unobstructed, clearly delineated, space that is 12 to 18 inches between walls and stored item, often referred to as 18 inch rule of sanitation.
- o Storage of food off the floor and easily movable by pallet jacks, forklifts, or other similar devices.

Self-inspections help to ensure a wholesale food warehouse is operating in accordance with its risk control plan (Anonymous, 2002). The measures put in place by a facility to control risks will go a long way to ensuring food safety and security. Efficient risk control system will assist in identifying problems more efficiently that need to be corrected, and customers should be assured that a comprehensive system is in place and the food they purchase is safe and unadulterated (Anonymous, 2002).

10.11 AUDIT/DOCUMENTATION AND RECORDS

For proper storage operation, a periodic audit of the whole system should be carried according to the Standard Operating Procedure, which is conducted according to good manufacturing practices/good hygienic practices (GMPs/GHPs). In order to provide an unbiased assessment of facility operations, try to ensure compliance with working risk control plan by any neutral outside source. All the appropriate records of storage, distribution, cleaning, and sanitation, pest control, and product recall should be kept and updated from time-to-time. Record must be retained for a period of one year or for the shelf life of the product, whichever is more. All cleaning and housekeeping within the warehouse must be thoroughly documented, and the records are to be available whenever prompted (Anonymous, 2015). There shall be a pest control program, documents such as layout, trending, and expectations.

Contract agreements shall be established, where applicable (Anonymous, 2017). There shall be written procedures assigning responsibility for sanitation and describing cleaning schedules, methods, equipment, materials to be used, and facilities to be cleaned in sufficient detail. Such written procedures shall be followed and the result shall be documented (Anonymous, 2002). Records that should be maintained must include:

- Risk control plan;
- Invoices and Shipping documents;
- Verification of approved sources;
- Temperature control logs or computer print outs;
- Thermometer calibration logs;
- Monthly facility self-inspection logs;
- Receiving/transportation/delivery logs;
- Training records;
- Cleaning logs and schedules;
- Disposition of salvaged, expired, damaged, or contaminated foods;
- Information and documentation for implementation of recalls;
- Water sample testing results;
- Pest control service receipts (it also includes assessment of environment as part of agreement);
- Material Safety Data Sheets for use of pesticides on-premises;
- Map of all pest control devices/bait stations;
- Weekly pest control inspection logs;
- Regulatory agency inspection reports/verification documentation.

The document could be in the form of record annexes for example (Anonymous, 2015):

- **Annex 1:** SOP on personal hygiene.
- **Annex 2:** Record daily inspection on personal hygiene.
- **Annex 3:** General health examination.
- **Annex 4:** Employee illness report form, record.
- **Annex 5:** Sample of pest control inspection, record.
- **Annex 6:** Sample of pest control monitoring, record.
- **Annex 7:** Sample SOP on pest control.
- **Annex 8:** SOP on cleaning of equipment and accessories.
- **Annex 9:** Cleaning schedule form.
- **Annex 10:** Cleaning record form.
- **Annex 11:** Cleaning inspection record.

Procedure and record of personnel hygiene for employees who perform manufacturing processes should be available. Thus, documentation and record-keeping are important in tracking a facility's operation (Anonymous, 2002). An efficient system needs to be in place to ensure records are being properly maintained as outlined in the plan. Records also need to be organized, stored in a secure location, and maintained for an extended period of time (Anonymous, 2002). One important concept to remember is, if you didn't record it, it didn't happen!

10.12 CONCLUSION

Fresh products require proper handling and special care throughout the entire supply chain. Complexities of regulatory compliance are becoming more challenging day by day. Any improperly stored fruit and vegetable can pose a health and safety hazard to both workers and customers, in addition to major loss in revenue it can bring pollution when all that spoiled food have to be thrown out. The goods stored in warehouses can bring irreversible damage, generally from infestation and contamination. Hence good sanitation and hygiene process is important in reducing losses during storage. Scientific storage and advances in developing various storage structures remains important. Effective Management of Sanitation and Food Safety is the decisive factor in reducing and eliminating adulterated and compromised food. Proper sanitation and hygiene process during storage is necessary in order to avoid contamination and to preserve the natural state of the product. Aside from adhering to regulatory requirements with regards to warehouse hygiene standards, it is equally important to prolong the product's shelf life and keep products in a healthy state for human consumption. All warehouses should follow the Hazard Analysis and Critical Control Point System (HACCP) as mentioned in the Food Safety Certification or ISO. Besides general cleaning, pest control management is also important in warehouses. Warehouses provide an abundant food source to rodents, insects, and birds. Hence, to pass business audits for licensing and accreditation, warehouse pest control is an important requirement. Accredited pest control technicians should visit the warehouses for scheduled inspections to identify and remove the earliest signs of a pest infestation before it spreads to other areas of your warehouse. This system should also follow in cleaning of warehouses. Measures such as regular inspection, good hygiene, careful attention, and maintenance to building design and better staff training will all contribute to maintain a clean, hygienic, and pest-free warehouse. Thus, warehousing solutions view sanitation and cleanliness as important ingredients for building good business practices that companies seek in prospective partnerships to best meet their warehousing needs.

KEYWORDS

- **cover and plinth storage**
- **first expire first out**
- **first in-first out**
- **good hygienic practices**
- **good manufacturing practices**
- **good warehousing practices**
- **hazard analysis and critical control point system**
- **integrated pest management**
- **pest control operators**

REFERENCES

Anonymous, (2002). *Wholesale Food Warehouse Risk Control Plan Workbook*. County of San Diego Department of Environmental Health Food and Housing Division. Retrieved from: https://www.sandiegocounty.gov/content/dam/sdc/deh/fhd/food/pdf/publications_wholesaleworkbook.pdf (accessed on 21 December 2020).

Anonymous, (2015). ASEAN Guidelines on GMP for traditional medicines/health supplement. *Chapter 4: Sanitation and Hygiene*. Retrieved from: http://asean.org/storage/2012/10/ASEAN-TMHS-GMP-Training-Chapter-4-Sanitation-and-Hygiene-FD.pdf (accessed on 21 December 2020).

Anonymous, (2015). *Food Grade Warehousing Standards and Guidelines*. Commercial Warehousing. https://www.commercialwarehousing.com/news/food-grade-warehousing-standards-and-guidelines/ (accessed on 21 December 2020).

Anonymous, (2015). *Food Safety in Distribution Centers and Warehouses*. Food First Blog. AIB International. Retrieved from: https://www.aibonline.org/Food-First-Blog/PostId/25/risky-business-food-safety-in-distribution-centers (accessed on 21 December 2020).

Anonymous, (2015). *The Warehouse Workforce*. Clean Middle East. Retrieved from: http://www.cleanmiddleeast.ae/articles/673/the-warehouse-workforce.html (accessed on 21 December 2020).

Anonymous, (2015). TNAU agritech portal. *Agricultural Marketing and Agri-Business: Agro-Processing*. Retrieved from: http://agritech.tnau.ac.in/agricultural_marketing/agrimark_storage%20and%20ware%20housing.html#ware (accessed on 21 December 2020).

Anonymous, (2016). *Six Things to Look for in a Food Grade Warehouse*. Kanban Logistics Blog. Retrieved from: www.kanbanlogistics.com/6-things-look-food-grade-warehouse/ (accessed on 21 December 2020).

Anonymous, (2016). *The Importance of Warehouse Sanitation and Cleanliness in Packaging and Warehousing Services*. Challenge Packaging Services Pvt. Ltd. Retrieved from: http://cpacks.com.au/the-importance-of-warehouse-sanitation-and-cleanliness-in-packaging-and-warehousing-services.html (accessed on 21 December 2020).

Anonymous, (2017). *Five Common Food Storage Mistakes for Warehouses*. The Shelving Guy Blog. Retrieved from: http://blog.shelving.com/warehouse/five-common-food-storage-mistakes-for-warehouses/ (accessed on 21 December 2020).

Anonymous, (2018). *Underground Storage of Grains*. Department of Primary Industries and regional developers, agriculture and food. Retrieved from: www.agric.wa.gov.au/underground-storage-grain (accessed on 21 December 2020).

Bodholt, & Diop, (1987). *Construction and Operation of Small Solid Wall Bins*. From FAO Agricultural services bulletin. ISBN: 92-5-102535-5. Retrieved from: http://agris.fao.org/agris-search (accessed on 21 December 2020).

Food Storage Warehouses, (2017). *Washington State Department of Agriculture*. Retrieved from: https://agr.wa.gov/departments/food-safety/food-safety/food-storage-warehouses (accessed on 21 December 2020).

Frontline Services Australia Pty Ltd (2015). *Storage Temperatures for Fruits and Vegetables*. Frontline Services Australia Pvt. Ltd. Retrieved from: https://www.frontlineservices.com.au/Frontline_Services/Storage_temperatures_for_fresh_produce.html (accessed on 21 December 2020).

Gangahar, P., & Tomar, K. P. S., (2009). *BFN-001: Introduction to Food Safety*. Indira Gandhi National Open University (IGNOU), ISBN: 978-81-266-3958-8.

Heaps, J. W., (2006). *Insect Management for Food Storage and Processing* (2nd edn. p. 21). Chapter 3.

Nijasure, S., (2013). *Safe Storage of Food in Warehouses*. Pest Control (India) Pvt. Ltd. Retrieved from: https://www.cleanindiajournal.com/safe-storage-of-food-in-warehouses/ (accessed on 21 December 2020).

Schuler, G. A., Nolan, M. P., Reynolds, A. E., & Hurst, W. C., (1999). Cleaning, sanitizing, and pest control in food processing, storage and service areas. University of Georgia College of agricultural and environmental sciences cooperative extension service. *Food Science*, 3–10. Bulletin 927. Retrieved from: https://seafood.oregonstate.edu/sites/agscid7/files/snic/cleaning-sanitizing-and-pest-control-in-food-processing-storage-and-service-areas.pdf (accessed on 21 December 2020).

CHAPTER 11

INSECT PEST INFESTATION DURING STORAGE OF FRUITS AND VEGETABLES

MIFFTHA YASEEN,[1] BASHARAT AHMAD BHAT,[2] JINKU BORA,[1] YASMEENA JAN,[1] MUNEEB MALIK,[1] and Z. R. A. A. AZAD[3]

[1]*Department of Food Technology, Jamia Hamdard, New Delhi–110062, India, Mobile: +91-9953903217, E-mail: miffthayaseen@gmail.com (M. Yaseen)*

[2]*Department of Life Sciences, Shiv Nadar University, Dadri, Uttar Pradesh–201314, India*

[3]*Department of Post Harvest Engineering and Technology, Aligarh Muslim University, Uttar Pradesh–202001, India*

ABSTRACT

As far as human nutrition is concerned, products of soil (fruits and vegetables (F&V)) are considered to be a delicious, appealing, and healthy diet. They are soft, fleshy, and are edible parts of the plant. 'F&V' is encompassed with essential components like vitamins, minerals, fiber, and phytochemicals having a wide range of health benefits. 'F&V' contains minimal quantities of fat, salt, and sugar. They are a rich source of dietary fiber, and a high intake of 'F&V' can help to dwindle obesity and perpetuate a healthy weight. A sedentary lifestyle requires food that's nutritious, energizing, and easy to eat on the go, like 'F&V.'

However, foods grown from the ground (F&V) are breathing tissues and profoundly perishable items which require ideal postharvest technologies with a specific end goal to keep up their storage stableness and expand the time span of usability. Developing purchaser interest for convenience in food production and consumption, including product form, packaging, quality conservation, and all year accessibility has been driving the F&V marketing framework to one with an expanded focus on value addition and cost minimization by streamlining the distribution. Quality and preservance of F&V depend upon the cultivar, pre-harvest practices, climacteric conditions, maturity at harvest, harvesting practices, and postharvest handling circumstances, making shelf-life prediction a troublesome assignment when contrasted with other food products.

Food commodities of animal and agricultural base are put away for future utilization and for trade purposes. Amid storage, the quality trait of the produce is influenced by the pest infestation besides ecological and environmental factors. The pests mainly include birds, rodents, insects, and microorganisms. Inputs including manpower and finances invested in the preparation of food commodities will go waste, unless the produce is protected from the degrading agents during storage. Postharvest deterioration can be controlled by diminishing the storage temperature and respiration rate by adjustment of the atmospheric conditions encompassing the product, which would enhance dependability and broaden time span of usability of the fresh produce. The main aim of the accompanying chapter is to focus around the diversified aspects of insect infestations causing the contamination and spoilage of perishable as well as 'dried fruit and vegetable' products, hence, by leading to the overall economical loss from the agricultural sector. This chapter also aims on the various detection and control techniques which can be taken so as to eradicate the loss of various nutrients right from the primary produce on the fields to the target population that is humans.

11.1 INTRODUCTION

India is an agricultural country. It is the second-largest producer of fruits in the world, just behind China. The agricultural base in India is the driving force for broad-based economic growth, but the main emphasis should be provided on reducing farm losses, improvement of storage conditions and, by processing of food, food products and other value-added products; which will impact export capacity in the global market level concerning different nations.

The increased globalization and flexibility of market prices, emphasized by the World Trade Organization (WTO), are creating new export markets for Indian agricultural products, including both fresh and processed food products (Deininger and Sur, 2006). One of the major challenges among them is for the agricultural sector, the demand for production of more food for the increasing population and to recompense for losses due to insect pests. During storage, food grains and other food products are critically disintegrated by insects and other pests.

Awareness among growers to identify and manage contamination and insect pests of various fruit crops are necessary for higher productivity as well as for quality produce. This requires long haul techniques to oversee dangers, for example, bugs, and infections following the standards of integrated pest management (IPM). Since there is an adjustment in utilization pattern of our nation, from grains to more broadened and nutritious eating routine of fruits and vegetables (F&V) and other value-added food products, dairy items, fish, meat, and poultry products. In India, this showed up in the development of the food-processing sector as a dawning industry.

India has an immense potential for increasing, generating employment opportunities, export earnings, and inspiriting agricultural economy. Since India positions second-biggest maker of leafy foods, fifth as far as creation, utilization, trade, and expected development. India is the biggest producer of milk, cereals, and in spite of the colossal generation of nourishment in India, we are unable to satisfy the fundamental needs of the Indian populace and of the underprivileged societies like having moderate, adequate, and healthy food. Indeed, even in today's situation, our nation is managing the issues of food inflation and food security. The administration of India has begun different activities like the Food Safety and Standards Act (FSSA) which was acquainted in 2006 to overcome the food safety issues and to give emphasis to safety standards and to safeguard food safety and health of the consumers. In these different exercises throughout the food dissemination chain, from essential production region through distribution area to retail and catering are covered.

The Food Safety and Standards Authority of India (FSSAI) were built under the FSSA (2006). It is a statuette body for framing science-based models for articles of nourishment and sustenance, which include food and food products with a specific end goal to control their manufacture, storage, dissemination, sale, and import, and to guarantee the accessibility of safe and healthy nutritious food for human utilization (FSSA, 2006).

In spite of the fact that India being the farming nation, shares the worldwide market almost around 1% just, however, there is expanding acknowledgment of cultivation from the nation when contrasted with alternate nations (Abhishek et al., 2014). Apart from huge speculation encouraged in by the private division, open segment (public sector) has likewise taken activities and with APEDA coming up with a couple of Centers for Perishable Cargoes and consolidated postharvest handling facilities have been set up in India.

The diversified climate of India assures accessibility of all assortments of fresh F&V. The creation of the products of the soil (F&V) in India is most astounding, just next to China. The reports of the National Horticulture Database disseminated by the National Horticulture Board, in the year 2014–15 reported that India created 86.602 million metric huge amounts of products of the soil. The colossal creation base offers India gigantic opportunities for export.

A total export of worth Rs. 10,369.96 crores/1,552.26 USD Millions which contained natural products (fruits) worth Rs. 4,448.08 crores/667.51 USD Millions and vegetables worth Rs.5,921.88 crores/884.75 USD Millions from India in the year 2016–2017. The nations like UAE, Bangladesh, Malaysia, Netherland, Sri Lanka, Nepal, the UK, Saudi Arabia, Pakistan, and Qatar are the primary targets for Indian fresh produce (Halder and Pati, 2011).

In the category of vegetables, India positions second underway of potatoes, onions, cauliflowers, cabbages, and brinjal. India positions first underway of ginger and okra amidst all vegetables. Among fruits, India positions first underway of Bananas (26.03%), Papayas (44.50%), and Mangoes (mangosteens, and guavas) (40.76%).

Despite the huge producer of the richest collection of mango cultivars, these had been banned by the European Union (EU) because the dispatches from India were infested with different sorts of fruit flies.

11.2 INSECT INFESTATIONS IN FRUITS AND VEGETABLES (F&V) INFESTATION

Infestation is the presence of an unusually large number of pests (insects, microorganisms animals) in a place, which cause harm (damage or disease) beyond the economic threshold level (Economic peak value is the maximum value of pest population that can be tolerated without economical losses). India, being the world's biggest maker of mangoes, develops almost around 15 million tons of the organic products, about 40% of the worldwide generation (British Broadcasting Corporation (BBC), 2014). Mango, alluded to as the "King of organic products," is a standout amongst the most critical industrially developed natural product in India and holds first rank among significant mango-delivering nations representing 50% of the world's creation. Major mango producing states of India are Uttar Pradesh, Bihar, Andhra Pradesh, Odisha, West Bengal, Maharashtra, Gujarat, Karnataka,

Kerala, and Tamil Nadu. Among these states, Uttar Pradesh ranks first in production with 3,841.00 thousand tons. According to the Agriculture and Processed Food Products Export Development Authority (APEDA, 2014), of the 55,600 metric huge amounts of mangoes traded by India in 2012–2013, just around 3,890 metric tons were sent out to nations of the EU. The EU forced transitory prohibition on imports of Alphonso mangoes from India.

The restriction also covered shipments of other Indian produce like vegetables, including bitter gourd, snake gourd, and eggplant, from May 2014 until December 2015 (BBC, 2014). The restriction was imposed after authorities in Brussels (Belgium) discovered 207 consignments infested with fruit flies.

11.3 TYPES OF INSECT INFESTATIONS IN FRUITS AND VEGETABLES (F&V)

The infestation by insects caused to F&V starts from the field itself, which continues till the fruits reach the consumer. Various types of insects cause damage to F&V, which are classified as in subsections.

11.3.1 FRUIT BORERS

Conopomorpha sinensis, which is also known as the lychee stem-end borer and the lychee fruit borer. It is the dominating pest in most of the seasons, and the host for *C. sinensis* and the related *C. litchiella* is lychee, and the major part attacked are preferably leaves and shoots, while other associated species *C. cramerella* is principally confined to rambutan and cocoa (Bradley, 1986).

Fruit borers (*C. sinensis*) lay yellow eggs which are scale-like and it generally measures about 0.4 x 0.2 mm long and develops on the natural product, and in addition on different parts of the plant-like new leaves and shoots in the wake of blossoming. Both lychee and longan are influenced. The hatching of eggs of C. sinensis takes place in three to five days, with the larva penetrating immediately into the whole plant including fruit, leaf, or shoot.

Fruits infested with fruit bores should be observed week after week to recognize eggs of *C. sinensis*. The eggs of *C. sinensis* are very small and generally not visible to the naked eye. If the infestation level of the fruit is 1–2%, the fruit must be picked and discarded.

Permethrin chemical has to be applied weekly, whenever, the pest becomes highly active. In the Taiwan Province of China, the application of different synthetics cypermethrin, deltamethrin, carbofuran or fenthion is prescribed in beginning times of natural product development with a specific end goal to maintain a strategic distance from harm later in the season. All the stages of leaf miner, the other fresh fruit infesting *Conopomorpha litchiella* Bradley, are similar in action to those of the fruit borer.

In a comparable way to organic product (fruit) borer, the female lays its eggs on new shoots of the plant, and the little; light-yellow eggs incubate for three to five days. The larva, which is newly has a creamy white appearance, and tunnels into the middle rib and veins of shoots and leaf blades. The insect named as *Cryptophlebia scirpophaga* Walsingham (*Argyroploce lepida* Butler) in India (Butani, 1977), is worldwide known as *Cryptophlebia ombrodelta* Loweris also a fruit borer species (Bradley, 1953).

It is widely distributed to Thailand, China, Japan, Taiwan Province of China, and Australia, but only in Australia, it is considered as a pest of concern. The grayish eggs of these species are oval and level with an intertwined surface, and measures around 1.0 x 0.8 mm. They are laid separately or in gatherings of up to 15 on the organic product (fruit) surface. The recently brought forth hatchling for the most part bolsters on the natural product (fruit) skin and after that channels into the seed of the fruit, which is eaten as whole fruit. A solitary hatchling may ruin a few natural products (fruits), if the organic product is little in size. However, they prefer fully-grown and mature coloring fruit with seeds that are larger in size.

11.3.2 FRUIT-PIERCING MOTHS

Eudocima (Othreis) *fullonia* (Clerck), *Eudocima salaminia* (Cramer) and *Eudocima jordani* (Holland) are the common fruit-piercing moths which are important throughout Asia, Australia, and the South Pacific. The host plants for the hatchlings of organic product (fruit) penetrating are the coral tree, Erythrina, and vines of the Menispermaceae.

The moths have an extensible tubular sucking organ known as proboscis that helps in boring a cavity into the outer covering of the fruit enabling the moths to imbibe the juice from the flesh. After the infestation of moth in the fruits, if contamination with yeasts and bacteria occurs, it causes complete damage to the fruit. Various species of *Drosophila* thus attract to the fermenting juice and hence, fastens the deterioration. Within a few days, a frothy fluid (mass of cells or exudates) permeates from the fruit and stains undamaged fruit close by.

11.3.3 LEAF-FEEDING CATERPILLARS

Oxides scrobiculate F. and Oxyodes tricolor Guen is the common among the leaf-feeding caterpillars known and is widely found in Thailand and Australia. *O. tricolor*, damages trees in southern Queensland (Australia), but in north it is not considered a pest. *Achaea Janata (L.)* which is usually known as castor oil looper in Australia, is a ravenous feeder. It frequently pervades trees in substantial numbers at the same time as *O. tricolor*. The caterpillars can cause acute defoliation. It is recommended to apply carbaryl when more than two young larvae are present per leaflet.

The efficiency of the sprays is improved by vibrating the tree to knock out the larvae onto the ground. If more than 40% of larvae are parasitized, sprays must not be necessary. When damaging populations of *O. tricolor* appear; *Bacillus thuringiensis Berliner (Bt),* endosulfan or methomyl are used as control agents in Australia.

11.3.4 LEAFROLLERS

Olethreutesperdulata Meyr., Platypeplus parabola (Meyrick), Adoxophyes cyrtosema Meyr. and *Homona coffearia Nietner* are the majorly known leaf rollers. These species of leaf rollers are widely found in Queensland, Australia, China, and India. These species of leaf rollers along with *Homona difficilis* are also found in lychee, longan, and rambutan in Thailand. One of the leaf roller known as orange fruit borer, *Isotenes miserana (Walker)* attacks flowers and fruits in Queensland. A minor pest *P. aprobola* found in China and India, predominantly assaults leaves and blooms of the plant. In Australia, it is present among a group of complex species which contribute to a significant loss of whole flowers. *A. cyrtosema* and *H. coffearia* additionally feed on entire plants in China. The waste caused by leaf rollers is endured in as much as it is constrained to the foliage. Methomyl or carbaryl is applied when 20% of leaf flushes are harmed, keeping in mind the end goal to limit harm to young trees or at basic times of leaf development in more seasoned trees. In India, phosphamidon, fenitrothion, or endosulfan is sprayed for heavy infestations and rolled leaves that contain larvae are removed manually by hand during light infestations.

11.3.5 BEETLE BORERS

Aristobia testudo (Voet) generally known as longicorn beetle, is a hazardous pest of lychee and longan in Guangdong (Zhang, 1997). The grown-up bugs show up from June to August, with one generation for each year. The females curl around branches and chew off about 10 mm strips of bark, with the eggs laid on the wound and covered with fluid mass of cells known as exudate. The hatching begins from late August and stays under the bark until January when they move down into the xylem and bore burrows up 60 cm long. The white-spotted longicorn beetle, *Anoplophora maculata (Thomson)*, has been found to have a one-year life cycle.

The adults usually grow up in spring and females lodge about twenty eggs individually into T-shaped incisions in the bark, 0.5 m above the soil surface. The larval period endures around ten months, and this development phenomenon has been found in Taiwan province of China. In Australia, the longicorn beetle, *Uracanthus cryptophagus,* causes comparable harm.

11.3.6 SCARAB BEETLES

Xylotrupes Gideon (Linnaeus), usually known as the elephant beetle, is among the basic known Scarab insects. The larvae develop in the soil or compost, where they nibble on plant roots, dust, and gravel. Later, they are drawn towards the fruit as they mellow, especially to those that have been damaged by birds. Chemical control generally gives unsatisfactory results. Evacuating the hatchlings physically is powerful in little size trees, however troublesome in extensive trees. Work is generally exorbitant in Australia, so this procedure adds fundamentally to expanding costs.

11.3.7 SOFT SCALES

Pulvinaria (Chlorpulvinaria) psidii (Maskell), for the most part recognized as the green shield scale, assaults trees in China, Australia, and India These soft scales are of the form of crawlers and are usually produced in spring by adult scales that infest the leaves and twigs. A portion of these crawlers move onto the entire plant-like in blooms and young fruits. The female scales are seldom confused for mealybugs because the egg masses that are embraced in sticky like wax filaments cover the ends of the scale. Delicate dark-colored scale, *Coccus hesperidum Linnaeus*, is usually found as a bug in Queensland, where synthetic substances have disrupted its parasitoids or it is protected by ants. The scales additionally deliver honeydew, which helps in the development of dingy shape on invaded products of the soil. These stained fruits aren't acknowledged in the commercial center.

Extreme invasions might be checked with methidathion, in spite of the fact that utilizing mineral oil is supported so compelling predators; the mealybug ladybird, *Cryptolaemus montrouzieri* Mulsant, and the green lacewing, *Mallada signata (Shneider*) are not influenced.

Mites Erinose mite, Acerialitchii (Keiffer), also called hairy mite, hairy spider, or dog-ear mite, are widely distributed throughout China, Taiwan Province of China, India, Pakistan, and Australia. Females lay eggs independently on the leaf surface, and the eggs measure around 0.032 mm in width, round fit as a fiddle and translucent white.

The adult mites measure about 0.13 mm long and bay pink in color. All phases of the mite have four legs, yet are very versatile, and move effectively from old leaves to pervade new flushes (Waite and McAlpine, 1992). At the point when pervasions are extreme, an expansive number of leaves is demolished, which for the most part causes no issues in set up trees, yet impact youthful plantations.

The issue is additionally serious, if the vermin shifts from leaves onto the creating blooms and fruits. Consequently, the natural product (fruit) set is harmed and disfigured, and in this way, the organic product ends up unmarketable. Different types of ruthless vermin, especially those from the Phytoseidae, have been recorded with A. litchi (Waite and Gerson, 1994).

11.3.8 GALL FLIES

The most common among the Gallflies is a leaf midge, *Dasyneura sp*., which has been found to be a major pest in China (Zhang, 1997). The hatchlings over-winter in the gallflies is created because of their nourishing pattern, and they, for the most part, pupate inside the dirt, with the grown-up flies starting the first of eight covering generations from March. The midges, for the most part, incline towards developing in soggy zones, finished shades and get dry in uncovered territories.

The eggs are laid in line on young leaves by the grown-ups. The hatchlings at that point make an opening in the leaf, causing 'watery dabs' that later turn into the —"galls." These galls turn dark colored and in the long run drop out, giving the leaf a shot-gap appearance (Zhang, 1997).

11.3.9 FRUIT FLIES

The commonly known fruit fly is the *Bactrocera tryoni (Froggatt).* The adult females lay their eggs through the outer skin of the fruit, often making use of cracks and wounds made by other pests. In spite of the fact that the eggs can incubate, the hatchlings of the fruit fly infrequently survive (De Villiers, 1992), likely as a result of the juice in mature natural product (fruit) suffocates them. The flies are regularly found in Australia, and other related species in Africa and Hawaii.

11.4 COMMON INSECTS OF PARTICULAR FRUITS

Some of the common insects of particular fruits are as follows:

1. **Banana Flower Thrips (*Thrips hawaiiensis*):** The scientific name is *Thrips hawaiiensis.* Little, thin-bodied dynamic insects are regularly observed proceeding onwards the external layer of young banana fruit, particularly close to the 'bell' or male end of the new bunch. Adult females (1 mm) are distinctly colored-dark orange and black-and are usually found under bracts or inside flowers of plant. Males (0.75 mm) are light yellowish straw-colored and are usually found on the outer layers of the bracts.

   Grown-up thrips have trademark wings; the straightforward wings have an edge of hairs encompassing the outside edge emerging in indistinguishable plane from the wing (Swaine and Corcoran, 1975). They are effortlessly observed with a 10X hand focal point.

   Banana bloom thrips are principle bugs in South East Queensland and northern New South Wales and minimum in northern Queensland. Thrips cause corky scab, which is essentially an issue in the drier banana-developing territories. In northern Queensland, thrips are seen dynamic for the entire year. Fruit obtained in winter or spring is usually the most affected, thus indicating that the activation period is greatest after the wet season.

   Fruit loss as well as spoilage is caused by feeding and Oviposition. Nourishing harm results in marginally raised regions on the fruit that are grayish-darker to dim silver at an introductory level. They create to shape the corky raised zones of dark-colored corky scab. Damage remains confined in most cases to the outer curve of the fruit, specifically near the soft end where the fruit finger joins the bunch stalk.

In serious pervasions, harm can move to different regions of the organic product. Lower hands (nearest to the male bloom) are most at infestation chance, however in extreme cases; harm can spread to cover the greater part of the group. Oviposition on young natural product (fruits) produces minute raised spots with a dull middle tip on the fruit surface. This damage has small economic importance since it becomes almost invisible as the fruit attains maturity.

2. **Banana Fruit Caterpillar (*Tiracola plagiata*):** *Tiracola plagiata* known as Banana fruit caterpillar has an extensive host range. The main commercial crops affected are the bananas and citrus family fruits. Adults generally range from medium to huge moths measuring 50–60 mm throughout the wings. The extreme gray forewings are dull gray-brown in color with dark brown V-fashioned vicinity at the fore margins. The hind wings are usually uniform with light brown-gray color and usually assault bunches on the outer edges of a plantation, close to scrub or rainforest. Larvae feed on each leaves and fruit. Large larvae generally feed deep inside the fruit while smaller, younger larvae feed on the rind of unripe fruit causing irregular-shaped brown colored patches of spoilage to outer fruit surfaces.

   The ratio of damage is intense and seen than that caused by the banana scab moth-damage that tends to be shallower and particular to the bottom of the fruit where it joins the bunch stalk. Because of their large size, one or two larvae are sufficient to destroy all the fruit on the bunch (Temperley, 1930).

3. **Mango Hopper (*Idioscopus clypealis, I. nitidulus, and Amritodus Atkinson*):** It is most dangerous and severe of all the mango pests and prevalent all over the country (India) causing heavy loss to mango cultivation. Though hopper populace exists during all the 12 months of year in mango orchards but sometimes it advances more between January to April on flowering flush The pest is observed during June-August on vegetative flush). Old, discarded, and closely planted orchards that are shady bushes and with high humid conditions favor their proliferation (Verghese, 1999).

4. **Black Scale (*Saissetia oleae*):** Black scale is the common name for *Saissetiaoleae* which causes main damage to olives, citrus, and gardenia. Grown-up females are circular formed dull dark colored to dark in shading, and measure around 2 mm. Little eggs are laid under

the female. These eggs bring forth into little, six-legged, cream white shaded 'crawlers,' generally in summer season.

After hatching the crawlers shift up the stems and usually stabilize along the veins of new leaves. The crawlers molt after about four weeks and then shift to the new stems and small branches. Here they will develop, creating defensive darker shells, and lay eggs.

Feeding damage is usually minimal; however, sooty mold commonly develops on the honeydew excretions from the outer scale, affecting photosynthesis (Flanders, 1942).

5. **Cluster Caterpillar (*Spodoptera litura*):** *Spodoptera litura* is the scientific name of Cluster caterpillar. It broadly damages F&V mainly strawberries, banana, sweet potato, tobacco, tomato, apple, cotton, cabbages, cauliflowers, many broad-leafed weeds.

   Grown-ups are grayish-brown moths with silvery speckle on the forewings; 16 mm in length and about 40 mm across the broadened wings. The hind-wings are crystalline in appearance. Eggs are laid in gatherings of up to 300 on the leaves by the night-flying moth and are covered with a matt of gray-brown hairs from the body of the female Young larvae are gregarious (hence the name cluster caterpillar) and feed on the axial surface of leaf, resulting a window effect. The young larvae are light green with blackheads. Conspicuous black triangles in a line are present along each side of the body of older hatchlings. The mature hatchlings pupate in the soil. The hatching of eggs takes 2 to 7 days, and the caterpillar stage lasts for about 2 to 6 weeks depending on temperatures.

   The life cycle that is from egg to adult, takes nearly about 30 days in warm weather and up to 8 weeks in colder conditions—moths (3 to 4.5 cm wingspan) lay batch of eggs on the underside of the leaf. Young larvae of *Spodoptera litura* feed in groups on either the top side or bottom side of leaves, leaving the opposite side intact. Large larvae remain solitary.

   Cluster caterpillar occurs throughout eastern Australia. Young larvae feed in very near groups and destroy one portion of the leaf, leaving the opposite side intact. Spoiled regions of leaf appear clear at first but turn brown immediately. On rare occasions, large-sized isolated larvae feed on fruit part resulting in the scarring of outer surface (Farrow et al., 1987).

6. **Orange Fruit Borer (*Isotenes miserana*):** *Isotenes miserana* generally known as orange fruit borer receives nourishment from many

plants including avocado, citrus, Feijoa, and Macadamia. The grownup moths are gentle grayish, speckled with minute spots which are brown in color, bell-shaped when at rest, with a wingspan of about 15–25 mm. They fly with a fluttering action, regularly in the course of night-time. The eggs mainly laid are scale-like, laid in clusters under leaves when fully grown the larvae measure about 24 mm long, brown on top and light grayish from beneath, with a dark brown head capsule and a pair of brown stripes throughout the body. They pupate inside the silken duvet which is fashioned whilst feeding. The pupae are brown or light green-brown in color, approximately measuring about 13 mm long and usually found on the dead leaves or foliage.

On hatching, the new young larva usually feeds on outer surface cells and immediately constructs a silken webbed like shelter. The one complete life cycle takes four to six weeks and further successive generations occur all the year-round. There are other generations in a year but the mode of activity is less in summer. All phases of the life cycle can be found in the winter season. The larvae suck, chews, and burrow inside the fruit slowly, just beneath the skin and often close the calyx.

Young and developed fruit can be damaged. Hatchlings roll onto flower buds and young leaves in cluster to form feeding shelters thereof. Larvae channel into fully mature and ripe fruit, causing it to fall, spoil, and finally decay, sometimes they often destroy new fruit in spring. New larvae may penetrate fruit just prior to harvest, and if not detected during packing, may lead the fruit to spoil and finally decaying during marketing (Faria et al., 1998).

7. **Pomegranate Fruit Borer (*Deudorix Isocrates*):** Pomegranate fruit borer is dispersed everywhere in India and Asia. It is the most common widespread, polyphagous, and spoilage-inflicting pest of pomegranate fruit. The mature adult female lays egg on main parts like flower, buds, and young fruits. On hatching, the caterpillar channels into the fruit and feeds itself on the pulp of the fruit. The fruits as a result of such spoilage then decay and drop off on the soil from the main plant.

The adult males are shiny bluish and brownish violet, and in case of females, a conspicuous orange patch on the forewings is seen. The damage of fruit borer is seen all the year-round. The female butterfly lays eggs on flowers, buds, and the calyx of developing fruits, after hatching; caterpillars bores into the fruit and feeds on the pulp (Murugan and Thirumurugan, 2001).

11.5 COMMON INSECTS OF PARTICULAR VEGETABLES

Some of the common insects of particular vegetables are as follows:

1. **Bean Blossom Thrips (*Megalurothrips usitatus*):** *Megalurothrips usitatus* usually called Bean Blossom Thrips are sometimes tiny cigar-shaped insects and measures up to 2 mm long. The adults are generally dark brown with a blood-red tinge on the surface.

   The nymphs or young are generally pale yellow to creamy white. Juvenile thrips are similar in form of adults; however, they don't possess wings. Adults lay their eggs within the cells of leaf tissue. After feeding, the under-developed ones drop from the plant and pupate within the soil. Generations are continuous, and their numbers are highest throughout the summer season. Temperatures of around 200°C favor reproduction and survival. Flower feeding causes twisting and curling of pods. The main vegetable affected is the French bean (Duff et al., 2014).
2. **Eggplant Caterpillar (*Scelio descordalis*):** *Scelio descordalis* is the scientific name given to eggplant caterpillar. The grown-up eggfruit caterpillar moths have yellowish-brown designed wings which measure about 25 mm wingspan. At the resting phase, they sit uniquely with the abdomen rolled towards the upper side. Eggs are tiny, flattened with a less longitudinal ridge. Formally, whitish-red markings appear as they start to mature. Larvae bore into the fruit. Young hatchlings are creamy-white while mature larvae are pink and about 20 mm long. The brown pupae are encased in tough whitish silken cocoons outside the fruit. Eggs are laid mainly on the calyx, hatch in 4–5 days at 25°C. Larvae channel into the fruit, and remain there until emerging to pupate. When the temperature is maintained at 25°C the larval stage takes about 10–17 days while as the pupal stage 6–14 days to complete. Eggfruit caterpillar remains active all year round in warm areas but has a winter diapauses in cold climates. Eggplant is the main commercial host, but it also sometimes attacks tomato, capsicum, and Pepino. *Solanaceous* weeds such as thorn apples and quena are hosts.

   Larvae attacks eggplant by deriving its nutrition within the fruit, making extensive channels that are usually filled with their waste material. Developed larvae leave a whole type (3–4 mm diameter) as they leave the fruit to pupae. Spoiled fruit will eventually break down and decay (Kay, 2010).

3. **Potato Moth (*Phthorimaea operculella*):** *Phthorimaea operculella* commonly known as potato moth attacks Pepino, potato, tomato, and tobacco, and few *Solanaceous* weeds which include thorn apple, false cape gooseberry and the nightshades.

   Both developed adult male and female moth's measures about 12 mm across the outspread wings, and have brownish-gray forewings with small dark distinctive markings. Adults have a fringe of fine hairs edges with the light yellow-cream hind-wings. The eggs are white and very small in size. The larva fully fed measures about 12 mm long and has a dark head. The body is grayish pink if the larva is in a tuber and dark green if it feeds on dead leaves.

   The pupa is dark brown and measures about 8 mm long. The larvae bore the whole plant. Young plants can be dead from topical point by boring larvae from that side. Larvae can bore through the calyx that is the soft end of the fruit, or at the point where two fruit or a leaf and fruit come together, causing spoilage beneath thereof. Damage is quite visible and tends to be more serious if other damage prone crops are grown nearby (Fenemore, 1988).
4. **Tomato Russet Mite (*Aculops lycopersici*):** *Aculops lycopersici* also generally named as Russet mites measures about 0.15–0.2 mm long and 0.05 mm wide. It can't be seen with naked eyes; therefore focusing lens is needed to see these mites. The crops mainly attacked are tomato, chili, and capsicum. Feeding by nymphs and adults of the tomato russet mite *Aculops lycopersici* results in loss of plant hairs, browning of the stem and finally decaying of lower leaves.

   The ripening of the fruits isn't uniform, and less mature fruits have light greenish with white spots. Infestations as well as spoilage are worst in peak summer season (Bailey et al., 1943).
5. **Root-Knot Nematodes (*Meloidogyne spp.*):** *Meloidogyne spp.* is minute, worm-like that are found mainly in soil. They usually have a wide range of host, and cause severe troubles in many annual and perennial crops. Tomatoes range among the most seriously affected fresh produce, with the nematodes resulting in various problems in all growing areas. Root-knot nematodes immature are very active, fibrous worms which usually measure about 0.5 mm long. They are too tiny to be seen with the naked eye. Many important fruit, vegetable, and ornamental crops are among the better hosts of these nematodes, like banana, cucurbit, grape, carnations, passion fruit, nectarine, kiwi fruit, chrysanthemum, pineapple, tomato, carrot, egg

fruit, strawberry, and many more. The members of the grass family are less procumbent to spoilage and injury than other plants to root-knot nematodes. Affected plants have a non-uniform shape and more often show symptoms of stunting, wilting, or chlorosis (pale color). The symptoms are of great concern when plants are infected soon after planting.

Mostly, nematode populace does not generally grow till late of the season and plants grow gradually until they reach maturity. Then they start to wilt and decay, dieback with flowering stage, retarding fruit maturity and fruit development. Beneath the ground, the signs of root-knot nematodes are quite unique. Lumps or galls measuring about 1 to 10 mm in diameter begin to develop all over the roots.

In serious damages, heavily galled roots may decay away, leaving a week root system with a few large galls (Bridge and Page, 1980).

11.6 INSECTS OF DRIED FRUITS AND VEGETABLES (F&V)

Wherever dried fruits are produced, whether in the Mediterranean Basin, South Africa, southern Australia, California, or Asia, their main insect pests are the same species. They have been distributed by commerce, possibly for several thousand years. The damage and loss caused by insects in dried fruits are challenging to assess. The loss of weight by volume from insect feeding is usually trivial. The most serious loss is in outer shape and quality, which decreases market value. In addition, the presence of bugs or any other foreign matter in dried fruit isn't liked by consumers.

The Food and Drug Administration (FDA) and the State Departments of Public Health insist that dried fruits must be prepared in a very hygienic manner, and regulations are strictly preceded. Further losses from insects in dried fruit include the extra charges of construction and maintenance of facilities for fumigation, the cost of fumigant, and the expense of applying them.

Insects are mainly responsible for the waste involved in culling out spoiled dried fruit, and the costs of screening and washing of fruit, general plant hygiene, and cold storage. Generally, five ways of storing fresh produce are used: drying, canning, curing, and salting, freezing, and common storage. The technique to be chosen depends on the type of crop, the desired quality, and the facilities possible for storage.

Dried 'F&V' are one of the approaches of preserving food for later use, and it will either be an alternate to canning or freezing, or advantage of these methods. Drying foods is simple, secure, and easy to understand, and drying eliminates the maximum of water from food so, the organisms with low water activity cannot grow and damage food. Although the dried foods serve an excellent source of fibers and more potassium to the consumers, yet these are also more calorie-rich and sweeter in taste. In dehydrated F&V insect infestation is one of the most prevailing household pest problems faced. In fact, all the dehydrated food products, including dry fruit and vegetable products, are vulnerable to pest infestations.

Dried food product pests are most commonly observed when they crawl or fly in the infested foods and most people find the spoiled products not suitable for consumption. The infestation of these dehydrated F&V leads to loss of both quantity and quality and hence market value. Adopting modern techniques of drying, occasional re-drying and well packaging and controlled storage facilities can lessen the losses of these cash crops.

The primary line of protection against insect pests should be well arrangement during production, next is careful harvesting techniques and preparation for marketing the produce and third is sorting out spoiled or decaying crops can cause contamination of rest healthy products. Some of the studies revealed that dried fruits like, Tamarind (*Tamarindus* India), Chinese date *Ziziphus Spina* chrisk (*Z. Mauritania*), balanites (*Balanites aegyptiaca*), were severely attacked while as infestation date palm (Phoenix dactylifera) was very less. Among the other dried vegetables like, chill pepper (*Capsicum frutescens* L.), Tomato (*Lycopersicon esculentus* Mills) and Bell-shaped pepper (*Capsicum annum* L.) had serious infestation while dried okra (*Abelmoschus esculentus* L. Moench) infestation was less during storage period of three month.

It was also found that Triboliumcastaneneum (Herbst, *Plodia interpunctella, Ephestia coutella,* and *Trogoderma granarium* infested all the dried produce during the study period. There were other insect species identified which were associated with dried F&V such as *Stegobium penicium* (L.), (Sitodrepapanicea) and *Teneboides mauritanicus* (L.). The destruction by pest of these stored dried produce led to loss of weight, value, and commercial value.

Adoption of new drying technologies, intermittent re-drying, use of good packaging and storage facilities engaged with excellent market structures can help to reduce the heavy losses of important cash crops. A number of beetles, moths, and to some extent flies also are the main pests of dried fruits, usually some of them may also be the major pests of cereals, nuts, and other stored crops (Simmons and Howard, 1975; Mound, 1989; Lewis, 1995; Okunade et al., 2001; Degri, 2007; Linda and Timothy, 2008).

There are other groups of beetles that are bothering to growers and they have distinctive features that make them serious pests, Dry fruit beetles are among members of the certain insect family known for its broad host range and ability to feed on various garden-fresh produce. These are very small pests, with tapered bodies and short antennae. Adults are normally black or brown, some bearing yellowish spots on their surface.

The larvae of the dry fruit beetle is similar to a small grub, that has a white body, tan head, and two spike-like structures emerging of its end. Adult beetles are vigorous feeders and have longer life in the adult stage, destroy dried fruits even more than their larvae. All of the major species, except saw-toothed grain beetle, infest food commodities mostly by flying or crawling into storage structures or by being carried along with the dried fruits. The dry products, especially fruits and nuts, are affected by damage causing organisms such as insects, mites.

Insect pests not only eat the food commodities and but also contaminate them with various insect parts; excreta and webbing. The spoiled commodities ultimately lose their freshness. These insects also distribute a variety of other microflora. They start attacking in the field during ripening of the fruits, storage, processing, transporting, and packing and finally in marketing mediums: Many other insect species attack different types of commodities such as oilseeds, cereals, and their related products; however, certain species are particular pests of dried fruits and nuts. Some are only limited to storage alone, while others occur in the field and in the storage area as well.

Following are the different types of insects of dried horticultural produce:

i. **Dried Fruit Moths:** Dry fruit moths that usually appear in dried fruits, the spoilage is done while these are in larvae phase. These insects survive and develop primarily spaces, although they are commonly brought into storage areas with infested commodities. Some of the common moths infesting stored F&V include *Ephestia castella* (Walker) *Cadra figulilella* (Gregson) (Raisin Moth), *Aphelia glares* (Zeller) (Dried Prune Moth) and *Amyelois transitella* (Walker) (Navel Orangeworm).

ii. **Beetles:** The beetles engaged with dried fruits show growth through four stages-egg, larva, pupa, and adult. The pace of development, duration of adult life, and number of eggs laid varies among the different species. As a rule, adult ones are active feeders. As some of them live for longer days in the adult stage than larva stage, the damage done is even severe than their larvae do.

All of the major species, sawtoothed grain beetle, can fly. They cause infestation commodities by crawling or flying into storage structures or by being carried in with incoming dried fruits. Some of the beetles damaging stored dried horticultural produce are as follows:

a. **Sawtoothed Grain Beetle (*Oryzaephilus surinamensis Linn.*):** The common name is saw-toothed grain beetle which belongs to family Silvanidae. The adult is chocolate brown in appearance, on each side of the edge of the body, it comprises of six teeth like shapes in front of the wings. It is a usual feeder and is abundantly found in stored raisins. Both of the forms larvae and adults are common pests infesting grains, nuts, cereal products, copra, botanical drugs, dried fruits, tobacco, candy, dried meat products, and unrefined sugar. In raisins stored for almost a year, even greater than one year, this insect can be present in a very huge in number. It is most occurring in spring; its population decline markedly during the peak summer hot months. It is a general feeder organism. Under certain conditions, the larvae are cannibalistic. Where saw-toothed grain beetles are numerous, in number of the Indian meal moth do not swell to peak levels (Back and Cotton, 1926).
b. **Dried fruit Beetle (*Carpophilus hemipterus Linn.*):** The adults are one eighth of an inch long and black in color with two amber-brown spots on each wing cover, with one near the tip and a smaller one at the outer side of the base. The fully grown larvae are pale white, head, and the anterior end of the body are amber-brown, less hairy, and have two visible spine-like projections at the posterior end, with two smaller ones in front of them. The two major pests are larvae and adults in case of ripening and drying figs that may be found in fruit dumps, decayed melons, stick-tight pomegranates, ground fallen peaches, citrus fruits, plums, cull figs, and moist raisins. They do not attack sound fruit, but prefer overripe, fermenting, and rotten fruit.

   This species thrives in fermenting grape pomace, a winery byproduct. When raisins are being made, damaged grapes, especially those with bunch rot, attract these beetles to the drying trays. Fruit that is very dry or far advanced in decay ceases to attract them. However, larvae that begin growth in overripe figs, for example, may continue their development after the fruit is fairly dry. Much waste ripe fruit falls to the soil under fruit trees and often squashes and cracks open. In date gardens, where frequent irrigation keeps the soil surface moist, dried fruit beetles find ideal breeding conditions in fallen waste dates (Vega, Dowd, and Nelson, 1994).

c. **Pineapple Beetles *Urophorus humeralis* (Fabricius):** These are commonly known as Pineapple beetles. The adults are lustrous black and are found in abundance in date gardens and feed on clusters of dates ripening on the palms, pineapple fields, decayed sugarcane seed pieces underground, and in waste grapefruit (Lindgren and Vincent, 1953).

d. **Darkling Beetles (*Blapstinus rufipes* Casey):** These are commonly called Darkling beetles. The adult is a dull grayish-black, somewhat flat in shape and these congregate on ripe figs that have dropped to the ground These beetles attack soft plants, girdle new bell pepper plants at the soil surface, and attack young plants of sugar beet, lima bean, and tomato (Davis, 1982).

e. **Cigarette Beetle (*Lasioderma serricorne* F.):** The adult beetle commonly known as Cigarette Beetle is light brown, small, oval, and about 3 mm long. The head and prothorax are bent downwards so as to give the insect a humped shape. It has been occasionally mislead with the drugstore beetle (*Stegobium paniceum*), but *L. serricorne* has serrate antennae and smooth elytra, whereas in *S. paniceum* the last 3 antennal sections are elongated and broad, forming a distinct club. The full-grown larvae measure about 4 mm in length, hairy, curved, and pupate in silken cocoons covered with small particles of foodstuffs. This beetle is thought to be the most damage-causing insect found in tobacco, but it also has the ranking of attacking a wider range of produce than any other storage pest.

Based on the reports of many investigators, plant materials damaged by the cigarette beetle include, areca nuts, aniseed, wheat crop, bamboo, biscuits, beans, cassava, chickpeas, cigarettes, cigars, drugs dried fruits coffee beans, cocoa beans, peanuts, coriander, copra, cottonseed (pre and postharvest), cumin, cottonseed meal, dried banana, dates, dried cabbage, licorice root, dried carrot, herbs, flax tow, herbarium specimens ginger, grain, juniper seed, insecticides containing pyrethrum, paprika, rice, rhubarb, seeds of various trees and plants, yeast, and spices.

It also breeds in leftover of animal, such as fishmeal, dried fish, dried insects, and meat meal. It has been found in damaging leather, stored wax, furniture inside suff, and may incidentally damage cloth, upholstery, plant-based products like books (Howe, 1957).

iii. **Flies:**

a. **Vinegar Flies (*Drosophila* species):** The vinegar flies, also called fruit flies or pomace flies, include species of *Drosophila* that are commonly seen wherever half spoiled or over matured fruit and vegetable trash accumulate.

These flies are mostly drowned towards fermenting fruit waste, melons, piles of peach and apricot pits, overripe grapes, tomatoes, or other fruits. They are common in and around tomato and fruit canneries. Cull fruit dumps lying on farms and on roadsides crowded with vinegar flies. By leaving rotting fruit and entering figs, these flies inoculate the ripening figs with yeast cells, and cause souring. They also add to the spread of bunch rot of grapes (Zhu et al., 2003).

b. **House Fly and Blow Flies:** Drying fruit is sometimes prone to contamination by flies that come from barns, cowsheds, or other nearby installations. These flies are not a primary pest of the fruit, but maybe a considerable disaster.

The most common species are the housefly, *Musca domestica* Linnaeus and several kinds of blowflies, *Phoenicia coeruleiviridis* (Macquart) *Phoenicia sericata* (Meigen), and *Orthellia caesarion* (Meigen) (Malik et al., 2007). Some of the insect pests on dry fruits are given in Table 11.1:

11.7 DETECTION

Detection and elimination of insects from fresh produce are major control actions for ensuring safe storage longevity, and food safety. Inspecting for insect-damaged fresh produce is labor-intensive and many infested fleshy fraction of the F&V may be undetected where an immature insect weather the flies or borers has not emerged from the core part of the fruit. Food inspectors at processing industries must be aware of the quantity of hidden insect infestation so that bulk of fresh produce with excessive destructions can be kept clean or diverted for other uses (Neethirajan et al., 2007).

Insect infestations in fresh produce cause both weight and quality losses and lesser export values. Insects not only eat fleshy part of fresh produce but also spoil it with their metabolic byproducts and other body fragments.

Insects produce moisture and heat due to their various metabolic activities, which in turn can lead to a growth range of microflora and thus resulting in the loss of whole fruit and vegetable. The heavily infested F&V and its products are not fit for human consumption.

TABLE 11.1 Common Insect Pests of Stored Dry Fruits and Nuts

Pest	Common Name	Scientific Name
Beetles	Darkling beetle	*Blapstinus rufipes* Casey
	Navel orange worm	*Amyelois transitella* Walker
	Hairy fungus beetle	*Typhaea stercorea* L.
	Dried fruit beetle	*Carpophilus hemipterus* L.
	Corn sap beetle	*Carpophilus dimidiatus* L.
	Confused sap beetle	*C. mutilate* Erichson
	Rust red flour beetle	*Tribolium castaneum* Herbst
	Drugstorbeetle	*Steegobium panaceum* L.
	Cigarette beetle	*Lasioderma serricorne* F.
	Saw-toothed grain beetle	*Oryzaephilus surinamensis* L.
	Merchant grain beetle	*Oryzaephilus mercator*
Moths	Rice moth	*Corcyra cephalonica* Stainton
	Raisin moth	*Ephestia figuliella* Gregson
	Codling moth	*Cydia pomonella* L.
	Dried prune moth	*Aphelia glares* Zeller
	Dried fruit moth	*Vitula serratilineella* Ragonot
	Almond moth	*Ephestia cautella* Walker
	Indian meal moth	*Plodia interpunctella* Hubner
	Oases dates moth/carob moth	*E. calidella* Guenee
	Tobacco moth	*E. elutella* Hubner

Infestation of either fruits or vegetables by pests such as insects during storage may make the fresh produce totally non-edible through associated microbial spoilage and contamination. From the commercial point of view, the fresh produce and its storage for a certain period of time where large volumes are held for long periods either for the seasonal availability throughout the year or for the processing purpose, the potential for loss is extremely high and are directly sum-able in financial terms (Neethirajan et al., 2007). External pest damage is the main target for the evaluation of outer surface defects in various fruits, for example, in jujube fruits. Pests can cause intense damage to fruits by altering their color and creating tiny holes on their surface. The presence of a few infested fruits in a shipment can make the entire shipment unmarketable and hence resulting in the less economic value in the market.

It is therefore important to identify fruits with insect damage before they are shipped to the market. Such identification will not only boost the salability of the product but also enhance or maintain its shelf life (Wang et al., 2011) Another example of insect infestation in terms of vegetables include soybean products pose potential hazard to consumers, thus making the food industry liable for economic losses (Huang et al., 2013).

The standards of quality for 'F&V' have been established in many of the countries to satisfy customers, among whom the awareness for clean, safe, and its products are increasing. With the increasing demand of end-users for high quality food, the pollutants such as insects, pests in postharvest produce must be minimized (Neethirajan et al., 2007). Before any control methods can be applied, the source of the infestation must be found. The physical ubiquity of the bugs (insects) is the most evident way for distinguishing zones of pervasion (infestation) and furthermore searches for old cast skins left by flour beetles. The existence of webbing is a simple means to identify items swarmed by Indian meal moth.

This dependably requires investigating all cracks and crevices where food debris might be, and checking inside containers of grains, beans, peas, flour, dried fruits, and other identical foods. There can be different sources like pet nourishment foods and birdseeds, which might become a source of invasion by insects. A sealed container may also get infested within and be sufficiently loose enough for an insect to getaway.

While examining food packages, only caterpillars may not be found, but rather silk webbing inside infested packages might also be present. Therefore, to overcome all the above-mentioned food infestation problems, a range of control measures are needed to be taken in order to ensure the wholesome and nutritious foods to the consumers and to boost the economic value as well.

11.8 CONTROL MEASURES

The various control measures adopted for controlling insect pest infestation are as follows:

i. Chemical method;
ii. Biological agents;

iii. Alternative approaches:
   a. Irradiation;
   b. Thermal method (low and high temperature).

11.8.1 CHEMICAL METHOD

This methodology involves the employment of chemicals for checking pest infestation. The chemicals commonly used are either in the disposition of insecticides or fumigants. In order to cut down losses due to insect pests, measures to control infestation are taken at every stage from harvest to storage and also during shipments. Fumigation plays an essential role in controlling insect pests during storage.

'Dried fruits and tree nuts' are disinfested chiefly with phosphine or methyl bromide as the scenario demands. Methyl bromide is offered in cylinders and in cans with or without a warning gas chloropicrin. Fumigation with methyl bromide can be carried out under normal atmospheric or at reduced pressures, for 4 to 24 hrs of exposure periods.

It is the apparent choice for controlling the dominant diapausing larvae of lepidopterous insect pests such as cydiapomonella in walnuts (Hartsell et al., 1991). The residue levels of inorganic bromide in the fumigated commodities must be within the prescribed tolerance limits. However, repetitive fumigations with methyl bromide are not recommended as chemical are been reported to transmit stench to tree nuts such as walnuts (Srinath and Ramchandani, 1978). Moreover, carcinogenic and ozone-depleting effects of methyl bromide has been reported. In recent years fumigation has come under severe criticism.

The fumigant is being eliminated in stages from 2000 onwards both in the developed and developing countries. Phosphine release out of metallic phosphide mixtures has been used for insect management when there is no time restriction. Because of its high vapor pressure and density is close to that of air and thus, phosphine gas gets distributed quickly and easily penetrates into the commodities during fumigation.

The gas is unstable and explosive at reduced pressures, and therefore, it is always used at ordinary atmospheric pressure only. Egg and pupal phases of the crawler insects are inordinately tolerant to phosphine. The efficiency of fumigant decreases at temperatures lower than 10° C. Also, at lower temperatures, the rate of respiration of insects decreases, and subsequently, the rate of developmental tolerant egg and pupal stages into larvae and adults are also prolonged.

Vail et al. (1993) reported that the fumigant exposure period length of 5 days is unfavorable for quarantine treatments. Because of the problems associated with phosphine and methyl bromide, extensive research has been conducted with the aim to find an alternative fumigant which is suitable for a range of stored products in combination with dry fruits and tree nut. Methyl iodide, carbonyl sulfide, and sulfuryl fluoride are reported to be suitable alternative fumigants specifically for treating dried fruits and nut (Zettler et al., 2000).

Carbonyl sulfide has been extensively studied for fumigations and has been considered as a suitable fumigant for treating stored products (Banks et al., 1993). However, there is an increasing demand to reduce the negative impacts of pest control methods on the environment and the humans as well. Also, with the increased concerns about the potential impact of pesticides on the well-being, the decrease in arable land per capita, and the evolution of pest complexes expected to be expanded, by atmosphere changes also add to change in plant safety practices.

Insecticides are still extensively used; however, more than 540 bug species are impervious to synthetic insecticides. Other downsides, of synthetic insecticides incorporate resurgence and outbreaks of secondary pests and harmful consequences on non-target organisms and to the population consuming the chemical treated foods (Banks et al., 1993). This situation develops a demand for alternative control techniques of insect pest infestation.

11.8.2 BIOLOGICAL AGENTS

Biological control can be characterized as the use of an organism to lessen the population density of another organism and in this way incorporates the control of animals, weeds, and diseases. There are three basic systems of biological pest management:

i. Classical;
ii. Augmentative; and
iii. Conservation control (Szabo, 1993; Van Lenteren, 2000).

Classical or inoculative control is utilized predominantly against 'exotic' pests that have turned out to be built in new nations or regions of the world. Moderately, small numbers (typically 1000) of a specific species of natural enemy are gathered from the nation or region of origin of the pest, 'inoculated' into the new environment, and permitted to develop the new level of control, which can be kept up over very long durations of time during the storage.

This kind of biological management has been most rewarding with perennial crops, where the long-lasting nature of the ecosystem enables the collaborations among the pest and natural rivals to become entirely recognized over a time period, for example, the successful use of the predatory ladybird *Rodolia cardinalis* for control of the unintentional presence of citrus pest *Icerya purchase* in Mediterranean Europe around 1900, and in Europe apple plantation presence of foreign wooly apple aphid, *Eriosoma lanigerum*, was controlled with the release of the parasitoid *Aphelinus mali* (Greathead, 1976).

Augmentation alludes to types of biological control in which natural predators are intermittently presented, and usually requires the merchandizing production of the released agents (Van Lenteren, 2000). Inundation includes the significant production and discharge of substantial amount of the control agent, such as *Trichogramma* egg parasitoids of various Lepidoptera pests including sugar cane borer *Diatraea saccharalis*, cotton bollworm *Heliothis virescens*, and European corn borer *Ostrinia nubilalis* (Bigler, 1986; Li, 1994; Smith, 1996; Van Lenteren and Bueno, 2003). Conservation control refers to the use of indigenous predators and parasitoids, usually against native pests.

Different measures are actualized to improve the abundance or activity of the natural enemies, which includes control of the crop micro-climate, formation of overwintering shelters (like 'scarab banks'), expanding the accessibility of alternative hosts and prey, and giving basic and essential food resources such as flowers for grown-up (adult) parasitoids and aphidophagous hoverflies (Landis et al., 2000; Wäckers, 2003; Winkler et al., 2005).

The principal chronicled evidence of biological management dates back to around 300 AD when predatory ants were utilized to control pests in citrus plantations (Van Lenteren, 2005; Van Lenteren and Godfray, 2005). In addition, in the 1880s, the cottony-cushion scale on citrus crops was controlled by a foreign ladybird (the Vedalia beetle-*R. cardinalis*) is broadly regarded as depicting the primary significant success of conventional biological control (DeBach, 1964) and dipteran parasitoid (*Cryptochaetum iceryae*), The primary achievement of seasonal inoculative biological(natural) control in protected cultivation included the glasshouse whitefly *Trialeurodes vaporariorum* and the parasitic wasp *Encarsia Formosa*.

The whitefly and other nursery green house pests (thrips, aphids, and spider mite) attack a variety of vegetable harvests (crops like cucumber, tomato, and peppers) and ornamental flowers. These pests lower yields, cuts down the market value of items that are sold on appearance, and give a breeding medium for secondary infections like in, sooty mold (*Cladosporium sphaerospermum*). Biological management of *T. vaporariorum* was recognized in UK before 1930 and ran effectively prior to the launch of DDT and few other organochlorine insecticides in the 1940s (Van Lenteren and Woets, 1988). However, the augmentation of resistance in the whitefly, together with other matters of concern about the overuse of pesticides, prompted the re-presentation of the natural control method in the late 1970s. For both *T. vaporariorum* and the glasshouse spider mite *Tetranychus urticae* controlled by the predatory mite *Phytoseiulus persimilis*, sophisticated administration plans have now been utilized, involving the development of the control agents, flexible release plans, augmentation of the pest groups to maintain natural enemy populations whenever required, and control of the climatic conditions to optimize control.

Currently, all nursery greenhouse vegetable pests can be persuaded with biological controlling agents (Van Lenteren, 2000). However, there are some limitations with the use of natural (biological) control agents. The fundamental drawback of biological control is that it is sluggish in curbing pest populations than most pesticides, as parasite organisms might take considerable days to get eliminated; and also, predators require a time frame to establish an economic level of pest suppression.

Development costs of natural control are occasionally depicted as 'high,' but these costs are much reduced than for the comparable synthesis, toxicological assessment and commercializing of a new pesticide, and considerable profits can be accomplished from bio-control with long term, effective natural enemies, of which *R. cardinalis*, *E. lopezi*, *E. Formosa* and *P. persimilis*.

The different species of *Cryptophlebia* are predated by their own particular complex of egg, larval, and pupal parasitoids; but, these don't generally keep borers beneath economic limits. Egg parasitoids, for example, *Trichogrammatoidea fulva* Nagaraja from India and *T. cryptophlebiae* from South Africa and Australia, offer the best prospects for organic control.

11.8.3 ALTERNATIVE APPROACH

11.8.3.1 IRRADIATION

The process of food irradiation includes the exposure of food products to radiant energy at a controlled amount which includes gamma rays or electron beams (ionizing radiation). Food irradiation is being used for over 50 years, especially in the EU. Food irradiation has proven to be an effective method of insect control and was first approved in 1987 by United States Food and Drug Administration (USFDA).

Biochemical and physiological changes leading to ripening in fruits make them more vulnerable to mechanical injury and attack by insects and rot producing fungal pathogens. Refrigeration, modified atmosphere, and elimination of ethylene can delay the ripening process, but has limitations as many of the tropical fruits show chilling injury or CO_2 injury.

Suppression of ripening in bananas, mangoes, papayas, and other subtropical, tropical, and temperate fruits have been observed when irradiated to low doses between 0.25 and 1.0 kGy (USFDA, 1987). The optimal dose may vary between fruits and also among some the fruit varieties. Irradiation increases the shelf life of fruits at non-chilling temperatures and while transporting over long distances (USFDA, 1987).

The combination of different physical treatments such as hot water dip, low dose irradiation can replace chemical fungicides for postharvest control of fungal rots (USFDA, 1987). Small-scale commercial application of irradiation to control spoilage losses of strawberries is being carried out increasingly in the USA since 1992. Irradiation at the dosages used for delay of ripening could also provide quarantine security against pests of fruit crops entering international trade.

Ionizing irradiation using cesium 137, cobalt 60, or linear accelerators is an efficient quarantine treatment that has a different measure of efficacy than all the other treatments used commercial level (Moy, 2005). Heat production by ionizing irradiation does not add to insect control. The higher death of pests (measure of efficacy) with irradiation, a higher dose is required than the amount of dose tolerated by fresh commodities.

Irradiation is an effective treatment for stopping the reproduction or causing sterility at lesser doses, and preventing the growth of exotic organism. Fruits to be shipped from Hawaii to the United States were irradiated before marketing.

Guavas were irradiated for infestation by Caribbean fruit fly in Florida. In Hawaii, an electron beam facility was built in for irradiating fruits. Sweet potatoes in Florida were irradiated for potato weevil before shipping to California.

South Africa began irradiating incoming grapes and few other fresh supplies that were considered a phytosanitary risk. Irradiation provides a sustainable, economical, and environmentally friendly treatment to control the development of a range of pests, spoilage organisms, and physiological processes in perishable crops.

National regulations, and international (Codex) standards are in place and public and food industry perceptions on the safety and benefits of the process are more positive. In view of the phasing out of the currently used postharvest chemical fumigants, irradiation either alone or in conjunction with other postharvest procedures can contribute towards the goals of achieving food security in developing and less developed countries by effectively reducing postharvest losses.

11.8.3.2 THERMAL METHOD (LOW AND HIGH TEMPERATURE)

In postharvest handling of agricultural as well as horticultural produce, temperature control is been broadly used to cut down the rate of degradation of produce caused by physiological processes, pathogens, and insects. Both high and low temperatures (Vincent et al., 2003) can be effective for the control of insects. The other contributing factors include temperature, rate of temperature change, and duration of exposure.

11.8.4 COLD STORAGE

Among the widely used treatments is the storage of fresh commodities at –0.6 to 3.3°C for 10 to 100 days, depending on the pest and temperature. It is used on a wide range of F&V (Follett, 2004). Low temperature conditions are suitable for a wide range of commodities, including many tropical fruits, such as apples stored for long periods of time, where low temperature could prove lethal to insects and thus increasing the marketing season.

Unlike most of the other methods, cold treatment can be useful after the commodities have been packaged and in slow transportation modes such as in ships. The main drawback is the length of the treatment period required. Freezing is mostly used for commodities that are usually processed, such as fruit pulp and fresh fruit and vegetables directed to the consumer market.

Freezing generally kills most insects in one day that are not in diapauses. Quick-freezing at temperatures of –150C typically kills diapausing insects (Follett, 2004).

11.8.5 HEATED AIR

Treatments using heated air were first used in 1929 during the fruit fly infestation in Florida. These treatments allow the exposure of commodities to air at temperatures range of 43 to 52°C from short to long time durations.

The treatment time can be reduced to one third if air is made to pass through the fruit load. The rate of treatment is dependent on temperature, size of individual commodity, speed of air through the load, compactness, and arrangement of the stack, and moisture content of the air.

These factors, in combination with the formation of heat shock proteins, modify the vulnerability of insect as well as the living commodity to heat, in terms of efficiency and product quality. Preconditioning by a mild heat treatment of fruits prior to pesticide exposure cuts down damage to product (Follett, 2004).

However, the same pretreatments make the pest extra tolerant to pesticidal treatment (Follett, 2004). Treatments by heated-air are commonly not well suited for temperate fruits. Treatments using dry air at high temperatures are used to treat agricultural commodities like grain, straw, meal, and dried plants.

11.8.6 HOT-WATER IMMERSION

Immersion in hot water at 43 to 55°C for a time period of few minutes to few hours is used to destroy a range of nematodes and arthropods on plant materials. Hot-water immersion is an easy, inexpensive, and speedy treatment. It is been used to disinfest all mangoes imported to the United States since 1987 (Sharp, 1987).

Several fruits, mostly those grown in temperate regions, get injured when immersed in hot water with a purpose to kill insects. The severity of injury is more in cases of regular heating. The tolerance to hot water immersion can be improved considerably by exposing such kind of fruits to 20 to 50°C for 0.6–72 h prior to the heat treatment (Sharp, 1987).

11.9 CONCLUSION

'F&V' are an important source of foreign commerce for the agricultural country like India and, therefore, need special attention regarding quality upkeep. Insect pests have been the significant deterrents for quality upkeep of these products. In this manner, control measures, particularly with reference to quarantine treatments, must be genuinely considered with the diapausing hatchlings (larvae) of moth pests and other hazardous pests as the target insects. Simultaneously, chemical residues and adverse effect on the products has to be taken into consideration. Walnut, for example, has been delicate to physical strategy such as irradiation and chemical techniques (phosphine and methyl bromide fumigations). In this regard, low temperature storage has been favored and is practiced for a few products however in future the role of controlled atmosphere (CA) and biological agents will be more critical in controlling insect pests in stored 'F&V.' The important pests of F&V include moths, beetles, and flies, among which the larvae of certain pyralid moths and the larvae and adults of nitidulid beetles are of significance. Fruits are even infested by certain species of insects before being harvested (when on trees). Other insects cause infestation and, ultimately, spoilage after complete drying, whether in storage or in packages. With the disparate habits of feeding and motility of pests such as beetles and moths, the control measures are to be taken over a larger area. Keeping in view of the economic concerns regarding the control of produce either fruits or vegetables, the various technological approaches are followed, which include the process of energy transfer to target insect pests and host plants to foster progress in equivocate practical situations. Laboratory examinations at present allow an accurate assessment of the efficiency-to-cost ratio for each novel technique to be used on a commercial scale for the production and supply of wholesome foods. Of particular concern are the widening technological and economic gaps among developed and developing countries. Although pastoral (agrarian) and entomological obstructions of temperate and tropical countries differ, the ratification of standards to accomplish the demands of global markets should not be at the expense of peasant agronomy. Specialized answers to manage insects should fit the financial and environmental existence of the nations like India.

KEYWORDS

- **Agricultural and Processed Food Products Export Development Authority**
- **balanced diet**
- **British Broadcasting Corporation**
- **degree Celsius**
- **European Union**
- **Food and Drug Administration**
- **fruits and vegetables**
- **infestation**

REFERENCES

Abhishek, R. U., Thippeswamy, S., Manjunath, K., & Mohana, D. C., (2014). Pest infestations and contaminants in foodstuffs: A major cause for the decline of India's contribution to the global food market. In: *Proceedings of Indian National Science Academy* (Vol. 80, No. 5, pp. 931–935).

Back, E. A., & Cotton, R. T., (1926). Biology of the saw-toothed grain beetle, *Oryzaephilus surinamensis* Linne. *Journal of Agriculture Research, 33*, 435–451.

Bailey, S. F., & Keifer, H. H., (1943). The tomato russet mite, *Phyllocoptes* destructor Keifer: Its present status. *Journal of Economic Entomology, 36*(5).

Banks, H. J., Desmarchelier, J. M., & Ren, Y., (1993). *Carbonyl Sulphide Fumigant and Method of Fumigation.* Intern. Publ. WO 93/13659.

Bigler, F., (1986). Mass production of *Trichogramma maidis* pint, et voeg. and its field application against *Ostrinia nubilalis* Hbn, in Switzerland. *Journal of Applied Entomology, 101*(1–5), 23–29.

Bradley, J. D., (1953). Some important species of the genus *Cryptophlebia Walsingham*, 1899, with descriptions of three new species (Lepidoptera: Olethreutidae). *Bulletin of Entomological Research, 43*, 679–689.

Bradley, J. D., (1986). Identity of the Southeast Asian cocoa moth, *Conopomorpha cramerella* (Snellen) (Lepidoptera: Gracillariidae), with descriptions of three allied new species. *Bulletin of Entomological Research, 76*(1), 41–51.

Bridge, J., & Page, S. L. J., (1980). Estimation of root-knot nematode infestation levels on roots using a rating chart. *International Journal of Pest Management, 26*(3), 296–298.

Butani, D. K., (1977). Pests of litchi in India and their control. *Fruits, 32*, 269–273.

Davis, J. C., (1982). New synonymy in *Blapstinus* discolor (Coleoptera, Tenebrionidae). *The Coleopterists' Bulletin*, 254-254.

De Villiers, E. A., (1992). Fruit fly. In: *The Cultivation of Litchis* (Vol. 425, pp. 56–58). Bulletin of the Agricultural Research Council of South Africa.

DeBach, P., (1964). *Biological Control of Insect Pests and Weeds*, 844.

Degri, M. M., & Zainab, J. A., (2013). A study of insect pest infestations on stored fruits and vegetables in the northeastern Nigeria. *International Journal of Science and Nature, 4*(4), 646–650.

Duff, J. D., Church, C. E., Healey, M. A., & Senior, L., (2014). Thrips incidence in green beans and the degree of damage caused. In: *XXIX International Horticultural Congress on Horticulture: Sustaining Lives, Livelihoods, and Landscapes* (Vol. 1105, pp. 19–26).

Faria, J. T., Arthur, V., Wiendl, T. A., & Wiendl, F. M., (1998). Gamma radiation effects on immature stages of the orange fruit borer, Ecdytolophaaurantiana (Lima). *Journal of Nuclear Agriculture and Biology, 27*(1), 52–56.

Farrow, R. A., & McDonald, G., (1987). Migration strategies and outbreaks of noctuid pests in Australia. *International Journal of Tropical Insect Science, 8*(4–6), 531–542.

Fenemore, P. G., (1988). Host-plant location and selection by adult potato moth, *Phthorimaea operculella* (Lepidoptera: Gelechiidae): A review. *Journal of Insect Physiology, 34*(3), 175–177.

Flanders, S. E., (1942). *Metaphycus helvolus*, an encyrtid parasite of the black scale. *Journal of Economic Entomology, 35*(5), 690–698.

Follett, P. A., (2004). Comparative effects of irradiation and heat quarantine treatments on the external appearance of lychee, longan, and rambutan. *Irradiation as a Phytosanitary Treatment of Food and Agricultural Commodities*, 163.

Greathead, D. J., (1976). *Mediterranean Fruit Fly, Olive Fruit Fly* (pp. 37–43). A review of biological control in western and southern Europe: Commonwealth institute of biological control, technical communication.

Halder, P., & Pati, S., (2011). A need for paradigm shift to improve supply chain management of fruits and vegetables in India. *Asian Journal of Agriculture and Rural Development, 1*(1), 1.

Hartsell, P. L., Tebbets, J. C., & Vail, P. V., (1991). Methyl bromide residues and desorption rates from unshelled walnuts fumigated with a quarantine treatment for codling moth (Lepidoptera: Tortricidae). *Journal of Economic Entomology, 84*(4), 1294–1297.

Howe, R. W., (1957). A laboratory study of the cigarette beetle, *Lasioderma serricorne* (F.) (Col., Anobiidae) with a critical review of the literature on its biology. *Bulletin of Entomological Research, 48*(1), 9–56.

Kay, I. R., (2010). Effect of constant temperature on the development of *Sceliodes corydalis* (Doubleday) (Lepidoptera: Crambidae) on eggplant. *Austral Entomology, 49*(4), 359–362.

Landis, D. A., Wratten, S. D., & Gurr, G. M., (2000). Habitat management to conserve natural enemies of arthropod pests in agriculture. *Annual Review of Entomology, 45*(1), 175–201.

Li, D. P., & Holdom, D. G., (1994). Effects of pesticides on growth and sporulation of *Metarhizium anisopliae* (Deuteromycotina: Hyphomycetes). *Journal of Invertebrate Pathology, 63*(2), 209–211.

Lindgren, D., & Vincent, L., (1953). Nitidulid beetles infesting California dates. *Hilgardia, 22*(2), 97–118.

Liu, X. D., & Lai, C. Q., (1998). Experiment on control of litchi stinkbug by using *Anastatus japonicus* Ashmead. *South China Fruits, 27*, 31.

Malik, A., Singh, N., & Satya, S., (2007). Housefly (*Musca domestica*): A review of control strategies for a challenging pest. *Journal of Environmental Science and Health part B, 42*(4), 453–469.

Moy, J. H., (2005). Tropical fruit irradiation-from research to commercial application. In: *International Symposium New Frontier of Irradiated Food and Non-Food Products* (pp. 22–23).

Murugan, M., & Thirumurugan, A., (2001). Ecobehavior of pomegranate fruit borer, *Deudorix Isocrates* (Fab.) [Lycaenidae: Lepidoptera] under orchard ecosystem. *Indian Journal of Plant Protection, 29*(1/2), 121–126.

Nanta, P., (1992). *Biological Control of Insect Pests* (p. 206). Biological Control Branch, Entomology and Zoology Division, Department of Agriculture, Bangkok, Thailand.

Sarwar, M., (2015). Protecting dried fruits and vegetables against insect pests' invasions during drying and storage. *American Journal of Marketing Research, 1*(3), 142–149.

Sharp, J. L., (1987). Status of hot water immersion quarantine treatment for *Tephritidae* immature in mangos. *Journal of Economic Entomology, 65*, 1372–1374.

Smith, S. M., (1996). Biological control with *Trichogramma*: Advances, successes, and potential of their use. *Annual Review of Entomology, 41*(1), 375–406.

Srinath, D., & Ramchandani, N. P., (1978). Investigations on fumigation of walnuts with methyl bromide. *Journal of Food Science and Technology, 15*(5).

Swaine, G., & Corcoran, R. J., (1975). Banana flower thrips and its relationship to corky scab damage of Cavendish bananas in southeast Queensland. *Queensland Journal of Agricultural and Animal Sciences, 32*(1), 79–89.

Szabo, P., Van, L. J. C., & Huisman, P. W. T., (1993). Development time, survival and fecundity of *Encarsia Formosa* on *Bemisia tabaci* and *Trialeurodes vaporariorum*. *WPRS Bulletin, 16*, 173–173.

Temperley, M. E., (1930). Life history notes on the banana fruit-eating caterpillar (Tiracolaplagiata Walk.). *Queensland Agricultural Journal, 33*(4).
U.S. Department of Agriculture, Animal, and Plant Health Inspection Service, (1987). Use of irradiation as a quarantine treatment for fresh fruits of papaya from Hawaii. *Fed. Regist., 52*, 292–296.
Umali-Deininger, D., & Sur, M., (2007). Food safety in a globalizing world: Opportunities and challenges for India. *Agricultural Economics, 37*(s1), 135–147.
Vail, P. V., Tebbet, J. S., Mackey, B. E., & Curtis, C. E., (1993). Quarantine treatments: A biological approach to decision-making for selected hosts of codling moth (Lepidoptera: Tortricidae). *Journal of Economic Entomology, 86*(1), 70–75.
Van, L. J. C., & Bueno, V. H., (2003). Augmentative biological control of arthropods in Latin America. *BioControl, 48*(2), 123–139.
Van, L. J. C., & Godfray, H. C. J., (2005). European science in the enlightenment and the discovery of the insect parasitoid life cycle in The Netherlands and Great Britain. *Biological Control, 32*(1), 12–24.
Van, L. J. C., & Woets, J. V., (1988). Biological and integrated pest control in greenhouses. *Annual Review of Entomology, 33*(1), 239–269.
Van, L. J. C., (2000). A greenhouse without pesticides: Fact or fantasy. *Crop Protection, 19*(6), 375–384.
Vega, F. E., Dowd, P. F., & Nelson, T. C., (1994). Susceptibility of dried fruit beetles (*Carpophilus hemipterus* L.; Coleoptera: Nitidulidae) to different *Steinernema* species (Nematoda: Rhabditida: Steinernematidae). *Journal of Invertebrate Pathology, 64*(3), 276, 277.
Verghese, A., (1999). Effect of imidacloprid, lambda-cyhalothrin and azadirachtin on the mango hopper, idioscopus niveosparsus (Leth.) (Homoptera: Cicadellidae). In: *VI International Symposium on Mango* (Vol. 509, pp. 733–736).
Vincent, C., Hallman, G., Panneton, B., & Fleurat-Lessard, F., (2003). Management of agricultural insects with physical control methods. *Annual Review of Entomology, 48*(1), 261–281.
Wackers, F. L., & Steppuhn, A., (2003). Characterizing nutritional state and food source use of parasitoids collected in fields with high and low nectar availability. *IOBC WPRS Bulletin, 26*(4), 203–208.
Waite, G. K., & Gerson, U., (1994). The predator guild associated with *Aceria litchi* (Acari: Eriophyidae) in Australia and China. *Entomophaga, 39*, 275–280.
Waite, G. K., & Huwer, R. K., (1998). Host plants and their role in the ecology of the fruit spotting bugs *Amblypelta nitida* Stål and *Amblypelta lutescens lutescens* (Distant) (Hemiptera: Coreidae). *Australian Journal of Entomology, 37*, 340–349.
Waite, G. K., (1990). Amblypelta spp. and green fruit drop in lychees. *Tropical Pest Management, 36*, 353–355.
Winkler, K., Wäckers, F. L., Stingli, A., & Van, L. J. C., (2005). *Plutella xylostella* (diamondback moth) and its parasitoid *Diadegma semiclausum* show different gustatory and longevity responses to a range of nectar and honeydew sugars. *Entomologia Experimentalis et Applicata, 115*(1), 187–192.
Zettler, J. L., & Arthur, F. H., (2000). Chemical control of stored product insects with fumigants and residual treatments. *Crop Protection, 19*(8–10), 577–582.
Zhu, J., Park, K. C., & Baker, T. C., (2003). Identification of odors from overripe mango that attract vinegar flies, Drosophila melanogaster. *Journal of Chemical Ecology, 29*(4), 899–909.

CHAPTER 12

POSTHARVEST DISEASES IN FRUITS AND VEGETABLES DURING STORAGE

MEETAKSH KAMBOJ,[1] NITIN CHAUHAN,[2] and PAWAS GOSWAMI[3]

[1]*School of Agriculture and Food, The University of Melbourne, Parkville, Victoria–3010, Australia*

[2]*Shaheed Rajguru College of Applied Sciences for Women, University of Delhi, Delhi, India*

[3]*Assistant Professor, Bhaskaracharya College of Applied Sciences, University of Delhi, Delhi, India, E-mail: pawasgoswami@gmail.com*

ABSTRACT

With rise in global population from last few decades, many developing countries are facing problem in meeting the rising food demand. Further, the losses in agricultural produce due to post-harvest diseases, especially in fruits and vegetables (F&V), increases this burden. Various pathogenic microorganisms, including, fungi (*Alternaria*, *Aspergillus*, *Botrytis* and *Rhizopus*, and *Mucor*), bacteria (*Erwinia*, *Pseudomonas*, *Xanthomonas*, and *Clavibacter*) and viruses (Tomato mosaic virus, Plum pox virus, and Potato Virus X) have been identified for major post-harvest losses in F&V. Several other factors such as, environment, physiology of post-harvested crops, inoculum level of pathogenic microorganisms and storage conditions directly influences the occurrence and spread of infection in the harvested F&V. The present chapter focuses on various factors responsible for post-harvest diseases and also uniquely summarizes the various post-harvest diseases of F&V caused by fungal, bacterial and viral species.

12.1 INTRODUCTION

The 21st century is regarded as an era of phenomenal advancements. We are witnessing improvements in the areas of technology, economy, and agriculture. With these improvements, especially in medical science, the world population is growing rapidly. The current world population has jumped up almost 150% from 3 billion in 1960 to 7.6 billion in 2018 (World Bank, 2018). With the increase in the population growth rate, especially in developing countries, and problems like deteriorating agriculture lands, declining soil fertility, incompetent farming, and above all, the shortage of storage facilities, developing countries are facing extreme scarcity of food. It was estimated that 1.3 billion tons of food get wasted annually due to various reasons (FAO, 2013). The biggest contributor to agricultural losses is postharvest diseases, with nearly 10–40% of agricultural produce being lost to it (Pallavi et al., 2014), imposing a maximum burden on developing countries. Postharvest diseases can be defined as diseases which develop on harvested parts of the plant due to the presence of microorganisms (Janisiewicz and Korsten, 2002). Most of the postharvest diseases of fruits and vegetables (F&V) are caused by fungi or bacteria. These microorganisms cause spoilage of both durable and perishable crops.

Various fungal species such *as Alternaria, Aspergillus, Botrytis, Rhizopus,* and *Mucor*, and many bacterial species, i.e., *Erwinia, Pseudomonas, Xanthomonas,* and *Clavibacter* were found to be responsible for postharvest losses. There are many ways by which an infection can occur, like direct penetration, natural openings, and through cracks present on the outer surfaces of F&V. Post-harvest diseases not only inflict great damage to the crop but also can result in various health hazards in animals and humans. The pathogens can cause partial or complete rotting or decaying of the crop, which results in a reduction of quantity and quality of the crops. Postharvest diseases have economic consequences also, as the disease can produce ill-effects on the crop, which makes it unmarketable; it can also result in the loss of an entire crop, leading to economic losses to the producer. In addition, the individual defective crop requires additional handling, which results in increased cost of production. There are various methods employed for the control of microbial contamination in F&V, i.e., physical methods, which include treatment with ionic radiations and high and low-temperature treatments. Further, chemical, and biological methods can also be employed to utilize chemicals and biological agents, respectively, for the control of infectious microbes.

12.2 INFECTION IN POSTHARVEST CROPS

Infection by postharvest pathogens can happen before, during, or after harvest. However, unlike in the case of pre-harvest pathogens, postharvest pathogens are incapable of penetrating the host crop. These pathogens enter the host through the mechanical injuries and thus are termed as wound pathogens (Barkai-Golan, 2001a). Even a microscopic injury is enough for the pathogen to penetrate the host. Mechanical harvesting can result in incision or scratches on the crops, which can be the entry points for the pathogens. Natural growth cracks in the crop can invite pathogen infection. Infection can also occur through cut stem by which microorganisms can enter the host and cause stem-end rots (Barkai-Golan, 2001a). Some fungus cause stem-end rot by systematically colonizing the stem, inflorescence, and fruit pedicel tissue without producing any symptoms of infection. This type of infection is called endophytic infection (Coates and Johnson, 1997). Penetration can also occur via physiological damages to the host, which is induced by various environmental stresses such as lack of oxygen, low temperatures and heat, etc. These factors increase the fruit or vegetable sensitivity and expose it to storage pathogens. It was also observed that a primary pathogen can ease the penetration process of successive pathogen (Barkai-Golan, 2001b). Snowdon (1990) observed that the prior infection in an apple by fungi, *Mucor, Gloeosporium,* and *Phytophthora* allows entry point for *Penicillium expansum*. Long storage times can lead to tissue decay, which can result in easy penetration of the host. During storage onions were found to be infected with basal rot disease caused by *Fusarium oxysporum* species (Coskuntuna and Ozer, 2008). Some postharvest pathogens infect the host before harvesting but remain quiescent until some changes in physiological or physical condition of the host triggers the pathogen to spread infection. Quiescent infections are very common in tropical fruits and are caused by both bacteria and fungi (Jarvis, 1994). Unlike most unripe F&V, which possess a mechanism for limiting quiescent pathogen aggression, postharvest fruits and vegetable are prone to the pathogen aggression due to the absence of any resistance mechanism. It has also been postulated that the inception of aggression is due to changes in physiological and physical properties of the host, particularly in the host cell wall (Barkai-Golan, 2001b; Narayanasamy, 2005).

Quiescent infections can result in symptomless, internal infections or invisible, nonexpanding lesions as in the case of ghost spot of tomato caused by *Botrytis cinerea* (Verhoeff, 1970). Most quiescent infection, if left undetected, can result in aggressive rots both in the field and in storage (Jarvis, 1994). Postharvest pathogens can also attack floral parts of the crop-giving rise to floral infections, which are most common amongst temperate fruits and can occur through any floral part like petal or stigma (Nabi et al., 2017).

12.3 FACTORS AFFECTING DISEASE DEVELOPMENT

A range of factors have been identified, affecting the disease development in post-harvested F&V. These factors include host physiology, inoculum levels, storage environment, etc. (Figure 12.1).

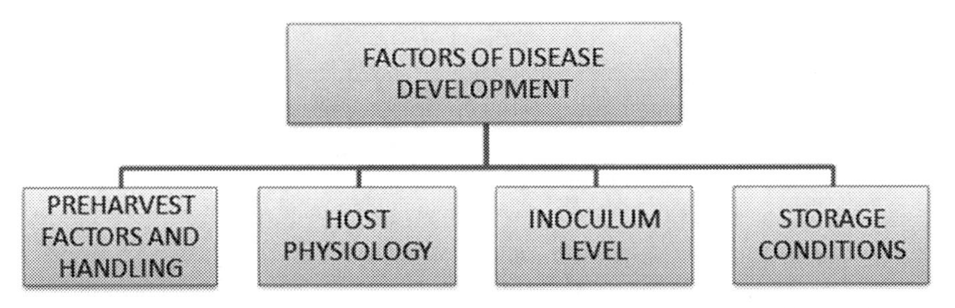

FIGURE 12.1 Important factors affecting the development of microbial diseases in post-harvested fruits and vegetables.

12.3.1 PRE-HARVEST FACTORS AND HANDLING

Many pre-harvest factors play an important role in disease development. The first factor is the health and quality of the planting material/seed. It is necessary that the planting material/seed is sterile and free from pathogens (Barkai-Golan, 2001c). In addition, with the advancement in agricultural science, disease-resistant seeds are available, which showed a substantial decrease in postharvest diseases caused by fungal and bacterial pathogens (Phipps and Park, 2002). Further, crop characteristics also play an important role in the quality and disease resistance of the postharvest produce. Crops like melon having thick skin are more resistant to pathogens and mechanical damage as compared to crops like strawberries, which are easily damaged due to thin outer covering (Barkai-Golan, 2001d).

Environmental factors such as temperature, humidity, rain, wind, etc., also play an important role during the developmental stages of agricultural crops, affecting disease development, quality of crops, and yield of produce

(Sharpies, 1984). Eden et al. (1996) found that high temperatures can increase the rate of flower development and senescence, which further increases the infection of *Botrytis cinerea.* Further, heavy winds, rains, and high humidity prior to or during harvesting can increase the risk of disease development. Moreover, wind and water also play an important role in the dispersal of pathogens and their spores. For example, spores of *Phytophthora* spp. germinate only when rainwater is available (Lapwood, 1977).

Studies have also highlighted that the higher plant density results in increased infection as a large number of host plants are available for the pathogen interception, further providing the ease of pathogen spread. Although pre-harvest antimicrobial sprays also limit the growth of pathogens, thereby affecting the disease development, but it was also reported that an increased plant density can decrease the potential of applied antimicrobials (Legard et al., 2000). Similarly, in another study, Eckert and Ogawa (1988) also reported that the chemical spray at pre and postharvest can result in the production of a strain that is resistant to the corresponding fungicide.

The method of harvesting also forms an important factor in postharvest disease development. It was noticed that the chances of infection are higher in mechanized harvesting than that of manual harvesting. During mechanized harvesting substantial damages are caused to the produce in the form of scratches and injuries, providing a gateway for pathogen entry. However, with proper training, manual harvesting can keep the damage to a minimum. Manual harvesting is preferred for vulnerable crops like strawberry or mango, whereas mechanized harvesting is used for less vulnerable crops or for the crops which has to be processed immediately like potatoes (Snowdon, 1990).

12.3.2 HOST PHYSIOLOGY

Physiology of host (post-harvested crops) can directly affect the process of disease development. To minimize the postharvest losses, it is important to understand the post-harvested physiology of the crops (Barkai-Golan, 2001c). Studies suggested that the attached or unattached parts of plants undergo senescence at their own pace. A study showed that the post-harvested plant organs (i.e., fruits) continue to respire and transpire by utilizing its food reserves and water, leading to senescence (Coates and Johnson, 1997).

The changes in the levels of pH, increasing softness, increase in the moisture content, absence or decrease in level of microbial resistant compounds (Phenolics), increased availability of compounds that induce pectolytic activity of the microbial enzymes makes the crop more susceptible to the pathogen

(Barkai-Golan, 2001e). Literature also highlighted that with ripening, the susceptibility of cell wall degradation increases as the links connecting the pectic substances to the cell wall breaks, leading to increased solubility of pectic substances, thereby softening the crop tissues (Eckert, 1978; Fabi et al., 2014). In this process, these pectic substances further induce and increase the activity of pathogen's pectolytic substances, helping in the penetration by pathogens.

12.3.3 MICROBIAL INOCULUM LEVELS

Inoculum concentration of pathogenic microorganism is directly proportional to the potential of the spread of infection. It was found that infection of tomato flower parts increased with an increased inoculum of *Botrytis cinerea* (Eden et al., 1996). Since in postharvest infections, pathogen can only enter through a wound, thus the development of infection depends on both the inoculum and the wounds present for penetration. Inoculum level on a crop is also related to the weather conditions, since microbial spores can be dispersed by winds and rains as in case of *Gloeosporium* and *Phytophthora* spp. (Barkai-Golan, 2001c). Further, with regular sanitation and antimicrobial spraying, inoculum levels of pathogenic microbes can be easily reduced in fields and storage houses.

12.3.4 STORAGE CONDITIONS

Storage conditions like temperature, relative humidity (RH), moisture content, sanitation of storeroom and overall storeroom atmosphere greatly affects the growth and infection caused by microorganisms.

12.3.4.1 TEMPERATURE STORAGE

Temperature affects not only the growth of microbe, but also the stored commodity. Therefore, the environmental temperature, the pathogens, and the hosts are actually inter-related. Different pathogens have different optimum growth temperatures. Most storage fungi require an optimum growth temperature of 20–25°C. Storage fungi have a range of minimal growth temperatures, with some fungi having a minimal growth temperature below 0°C, causing spoilage of even refrigerated F&V. The fungus, *Botrytis cinerea,* which has a minimal growth temperature of –2°C cause infection in celery, lettuce, and other refrigerated vegetables. Many pathogens have a minimal growth temperature of 9–11°C and their infection can be prevented by storing the produce at low temperatures. However, it is important to note that the storage temperature affects the produce too. Exposure of host to unfavorable temperatures can result in tissue damage, providing wounded or damaged sites for the pathogen entry, further enhancing the rate of infection (Barkai-Golan, 2001c).

12.3.4.2 RELATIVE HUMIDITY (RH) AND MOISTURE CONTENT

A high RH is maintained to prevent dehydration and weight loss in F&V. A report highlighted that the high humidity environment (RH> 95%) is needed to store fresh fruits which provides protection from shriveling and turgidity loss (Tian, 2007). However, reports also suggested that the high moisture content also provides the optimal conditions for the growth of various fungal species. As per study, high humidity in association with optimal temperature promotes the development of infection (Hollier and King, 1985). Therefore, maintaining the appropriate RH and preventing the growth of harmful microbes is a cumbersome affair.

12.3.4.3 STOREROOM ENVIRONMENT

The irregular sanitation of storeroom, presence of pests/rodents, and gas composition of the storeroom, all contributes to the development of disease. Gases present in storeroom directly affect the ripening and senescence processes of the crop, which in turn promotes the decaying of produce (Saltveit, 1999). The presence of pest can cause physical damage to the crops, thereby deteriorating the quality of the crop and also provides an entry point for pathogens. A proper and regular sanitation of storage room is very important to prevent disease development by reducing the pathogenic microbes present in the environment.

12.4 POST-HARVEST DISEASES AND CAUSAL ORGANISMS

The microorganisms which are involved in postharvest diseases belong to Fungi, Bacteria, and Viruses. Bacterial postharvest diseases are more predominant in vegetables, while fungi infect both F&V. Viral infections affect vegetables more than fruits. Only about 40 species of pathogens naturally attack harvested F&V and each fruit and vegetable have their own typical pathogen (Barkai-Golan, 2001d).

12.4.1 BACTERIAL DISEASES

Bacteria are primarily responsible for infections in vegetables. *Erwinia, Pseudomonas, Xanthomonas,* and *Clavibacter* are the major postharvest disease-inducing genera of bacteria. Table 12.1 highlights various types of bacterial species involved in causing diseases in different hosts. Bacterial pathogens primarily give rise to soft rots of the vegetables. The ability to secrete cell wall degrading (pectate) enzymes (lyase and hydrolase) is essential for the growth of bacterial pathogens. Bacteria are not able to penetrate the crop unless a wound is present, as shown in the case of *Pseudomonas cepacia,* where it could not infect intact onion, but caused infection in a wounded onion (Kawamoto and Lorbeer, 1974). Bacteria after entering through the wound cause maceration of parenchymatic tissue of the host, resulting in lesions and rots.

Erwinia carotovora causes diseases in various vegetables. *Erwinia carotovora* subsp. *carotovora* is a ubiquitous organism and cause blackleg and aerial stem rot disease of potato (Molina and Harrison, 1980). The optimal temperature of infection by *Erwinia carotovora* subsp. *carotovora* is around 25°C (Perombelon and Salmond, 1995). Being ubiquitous, it has a large number of hosts and can survive in soil and plant debris. It also has the ability to multiply within the soil, and any wound caused during harvesting or by insects is a potential entry site for it (Narayanasamy, 2005). In contrast to it, *Erwinia carotovora* subsp. *atroseptica* is limited to temperate regions and was found to be responsible for stem infection of potato. Its optimal infection temperature is around 15–20°C (Perombelon and Salmond, 1995). Infection of *Erwinia carotovora* also depends upon the host factors. It was found that high moisture content and high use of nitrogen fertilizer increased the susceptibility of Capsicum to *Erwinia carotovora* (Coplin, 1980).

Infection of *Pseudomonas cepacia,* which causes sour skin disease of onion increases at acidic pH due to the increase in activity of macerating enzyme, polygalacturonase (PG) (Kawamoto and Lorbeer, 1974). *Xanthomonas campestris* causes bacterial spot on capsicum. Its infection is favored by high humidity along with an optimal temperature of 20°C (Diab et al., 1982).

TABLE 12.1 Bacterial Diseases in Fruits and Vegetables

Fruits/Vegetables	Disease and Symptoms	Bacterial Pathogens	References
Tomatoes	Bacterial speck-necrotic lesion, black/brown spots on leaves and stem	*Pseudomonas syringae* pv. *tomato*	Mensi et al. (2018)
Carrots, tomatoes, onion, beans, sweet potatoes, capsicum, Avocado	Bacterial Soft Rot-Fleshy stored organs of plants turned into soft liquid mush	*Erwinia* sp., *Pseudomonas* sp., *Xanthomonas campestris*	Singh and Sharma (2018)
Apple	Fire blight-Leaf and shoot blight	*Erwinia amylovora*	Myung et al. (2016)
Kiwi	Bacterial canker-Canker formation	*Pseudomonas syringae* pv. *actinidiae*	Templeton et al. (2015)
Orange and lemon	Bacterial blast and black pit-Spotted and blistered fruits, black-discoloration of leaves	*Pseudomonas syringae* pv. *syringae*	Abdellatif et al. (2015)
Tomatoes	Bacterial canker-Cankers formed on stem, Yellow-brown vascular discoloration	*Clavibacter michiganensis* subsp. *michiganensis*	De León et al. (2011)
Mango	Bacterial black spot-Fruit lesion	*Xanthomonas campestris*	Manicom (1986)
Banana	Banana Xanthomonas wilt-Wilting and yellowing of leaves, excretion of a yellowish bacterial ooze	*Xanthomonas vasicola* pv. *musacearum*	Biruma et al. (2007)
Papaya	Bacterial leaf spot-Discoloration and distortion in leaves	*Pseudomonas caricapapayae*	Hasan et al. (2018)
Grapes	Bacterial leaf spot-Dark brown/black lesions on leaves	*Xanthomonas campestris pv. viticola*	Kamble et al. (2017)

Some bacterial pathogens are transmitted easily, but some infections only spread if a particular criterion is met. *Clavibacter michiganensis* subsp. *sepedonicus* which causes ring rot disease of potato spreads easily via irrigation water and further infecting other potatoes present in the water (Van der Wolf and Van Beckhoven, 2004). *Acidovorax avenae* subsp. *citrulli*, which causes watermelon fruit blotch didn't transmit to other fruits even when in contact with diseased fruit, albeit for a short period (Rushing et al., 1999). Therefore, disease development in the host by bacteria depends not only on host and pathogen factors, but also on environmental condition of storage.

12.4.2 FUNGAL DISEASES

Fungal pathogens cause prominent diseases both in vegetables and fruits (Table 12.2). They are the main cause of spoilage of fruits. The growth of fungal pathogen on F&V depends on the environmental conditions, host factors, and pathogen factors. Pathogen has to overcome host defenses and further induce certain enzymes and metabolites which aid in host tissue maceration and release of nutrients for its survival. Fungal pathogens can have both wide and narrow host range. The fungal pathogens cause both quantitative and qualitative losses to the crop. Fungal pathogens are easily transmitted and can perpetuate even in soil and crop debris for a long time by releasing various fungal spores. These spores cover the surface of vegetables and fruits. The fungal spores are resistant in nature and are easily dispersed off by wind, water, rain, or insects and have the potential to infect crops. Mostly, postharvest fungal spores are unable to penetrate the host and need a wound entry. When provided optimal conditions, these spores can germinate and cause infection in the host.

Mechanical injuries and chill injuries predispose fruits and vegetable to fungal infections. Saprophytic fungi like *Fusarium* and *Geotrichum* spp. can invade under appropriate condition for opportunistic invasion (Narayanasamy, 2005). Unlike bacteria, fungi not only affect the quality of the crop, but also produce toxins which make the consumption of such crops hazardous. Species of *Aspergillus, Alternaria,* and *Penicillium* genera are major mycotoxin secreting species.

Temperature influences the growth of fungi over host surfaces. Optimal temperature increases the development of postharvest diseases. Different fungi have different growth temperature, e.g., *Aspergillus niger* causing black rot and *Monilinia fructigena* which infects apple have optimum temperature of 35°C and 23–25°C, respectively (Mandal, 1981; Xu et

TABLE 12.2 Fungal Diseases in Fruits and Vegetables

Fruits/Vegetables	Disease and Symptoms	Fungal Pathogens	References
Strawberry	Rhizopus Rot-Water soaked, light brown fruits covered with fluffy mycelium	*Rhizopus stolonifer*	Lin et al. (2017)
Oranges	Black Rot-Atypical (black) coloration in fruits	*Alternaria alternata*	Soylu and Kose (2015)
Apple, kiwi, raspberry, pomegranate, blueberry, peach, sweet cherry	Gray Mold-Grayish colored mushy spots on fruits and leaves	*Botrytis cinerea*	Romanazzi et al. (2016)
Potatoes	Black Scurf-Irregular shaped tuber with black patches	*Rhizoctonia solani*	Hussain et al. (2017)
Carrots	White mold-White mycelial growth	*Sclerotinia sclerotiorum*	Ojaghian et al. (2019)
Eggplant	Fusarium rot-wilted foliage, brown discolored stem, stunted plant	*Fusarium solani*	Li et al. (2017)
Apples	Blue mold-blue pustules, fruit decay	*Penicillium expansum*	Da Rocha Neto et al. (2019)
Papaya	Stem end rot-Yellowing, drooping leaves	*Fusarium falciforme*	Gupta et al. (2019)
Grapes	Bunch rot-white to dark brown discoloration with fluffy fungal growth	*Mucor fragilis*	Ghuffar et al. (2018)
Onions	Basal rot disease-pinkish, brown rot covered with mycelium	*Fusarium oxysporum*	Coşkuntuna and Ozer (2008)

al., 2001). Generally, disease development recedes below 5°C as can be seen in the case of *Rhizopus stolonifer* (Dasgupta and Mandal, 1989). Thus, storage of crops at low temperature can help in controlling fungal infections. Further, RH also plays an important role in disease development. Fungal pathogens grow well in moist environment and studies showed that direct sunlight reduces the survival of the spores (Rotem et al., 1985). Interaction between RH and temperature is important in spore germination and growth of pathogen. Different pathogens have their own optimal combination of RH-Temperature for growth. *Colletotrichum gloeosporioides* which causes mango anthracnose disease grows at RH of 70–87% with temperature between 28 and 34°C, while some pathogens like *Helminthosporium solani* flourished at warm temperatures and RH of higher than 95% (Banik et al., 1998). Tomato fruit are susceptible to both pre and postharvest pathogens. Several fungal pathogens infect tomato fruit and cause disease under different environmental conditions. For example, *Alternaria alternata* causes disease at a RH of 100% and between temperatures of 21–30°C (Pearson and Hall, 1975), while *Rhizopus stolonifer* causes infection at 80% RH and at temperatures of 5–25C (Akhtar et al., 1999; Silveira et al., 2000).

Host factors also play an important role in fungal infection. Some hosts have disease resistance mechanisms and cannot be infected easily. For example, the R gene in potato imparts resistance against *Phytophthora infestans.* Many hosts possess antifungal compounds and toxins against fungal pathogens. For example, Onions with red scales produce phenolic compounds like catechol and protocatechuic acid, which prevents the infection of *Colletotrichum circinanas.* In the case of tomato, it was observed that unripe tomatoes are less susceptible to *Botrytis cinerea* due to presence of tomatine which act as toxin against the pathogen. However, the absence of this toxic glycoalkaloid in ripe fruit made it more susceptible to the fungal pathogen (Verhoff and Liem, 1975).

12.4.3 VIRAL DISEASES

Viral pathogens primarily infect vegetables, but studies also suggested infections in various fruits. Viruses are responsible for reduced growth and low yield. Tobacco mosaic virus (TMV), tomato mosaic virus (ToMV), plum pox virus (PPV), and Potato Virus X are all responsible for considerable crop losses. Table 12.3 summarizes various viral diseases affecting a variety of F&V. Viruses are very stable and can maintain infectivity in plant debris or even in dry soil for a considerable time period. The infected crops are the main

TABLE 12.3 Viral Diseases in Fruits and Vegetables

Fruits/Vegetables	Disease and Symptoms	Viral Pathogens	References
Tomatoes	Tomato leaf curl-Leaf curling, crinkling, yellowing, and mottling	*Tomato leaf curl New Delhi virus* (ToLCNDV)	Moriones et al. (2017)
Strawberry	Pallidosis-Reduce runner production and affects roots	*Strawberry pallidosis associated virus* (SPaV)	Tzanetakis et al. (2004)
Blackberry	Blackberry yellow vein-vein clearing, yellow mottling, ringspots in plants	*Blackberry yellow vein associated virus* (BYVaV)	Martin et al. (2003)
Melon	Cucurbit yellow stunting disorder-yellowing and stunting on melon	*Cucurbit yellow stunting disorder virus* (CYSDV)	Kao et al. (2000)
Lettuce	Lettuce chlorosis-Interveinal yellowing, stunting, rolling of leaves	*Lettuce chlorosis virus* (LCV)	Duffus et al. (1996)
Plum, peaches and apricot	Plum pox (Sharka)-mild to severe chlorotic and necrotic rings	*Plum pox virus* (PPV)	Collum et al. (2019)

transmission source. Infection can be spread through leaf contact or handling as in the case of ToMV (Broadbent, 1965) or by a vector as in the case of tomato spotted wilt virus (TSWV). PPV and potato viruses are transmitted by a variety of aphid species. Two different viruses can infect in conjunction and increase the severity of disease on the host. Disease development in host depends on environmental factors as well. While viruses can be stable at both low and high temperature, some viruses like TSWV get inactivated at temperatures of 45–50°C. Virus can be host-specific or can have a wide range of hosts. For example, TSWV infects 163 different species of plants (Francki and Hatta, 1981). Studies showed that different types of whiteflies are responsible for transmission of criniviruses, thereby posing threats to F&V, i.e., TICV (tomato infectious chlorosis virus) and ToCV (tomato chlorosis virus) affects tomato production. Further, whitefly transmitted SPaV (strawberry pallidosis associated virus) infects mainly strawberries and BPYV (beet pseudo-yellows virus) infects cucumber and pumpkin. Both TICV and ToCV highlighted interveinal yellowing and thickening of infected leaves, with decreased fruit size and early senescence, whereas SPaV and BPYV were found to highlight red coloration in older leaves with stunted roots, inhibiting plant development (Wintermantel, 2004). Studies have also highlighted that viruses have a wide range of hosts for infection.

12.5 CONCLUSION

Postharvest diseases are primarily caused by fungal and bacterial pathogens. Recent studies have also reported the role of various viral pathogens to impact the production of F&V. Disease development depends on the environmental, host, and various pathogen factors. Various bacterial species such as *Erwinia, Pseudomonas, Xanthomonas,* and *Clavibacter* and fungal species such as *Aspergillus, Alternaria, Rhizopus,* and *Penicillium* are found to be associated with diseases in post-harvested vegetables and fruits. The knowledge of disease development factors delineates the storage conditions and management, thereby minimizing the postharvest diseases. Although use of chemicals have controlled postharvest losses, but its heavy usage have also created adverse effects to environment and health, producing chemophobia among farmers and end-users. Various alternative approaches are booming from the last few years like biological methods, which help to control infection by utilizing plant extracts, essential oils, and antagonistic microorganisms. Moreover, effective control was also observed by using physical methods including heat treatment/cold treatment, low-pressure storage, and ionizing radiations and advanced packing system (where the packing material is coated with antimicrobial agent). Further, an integrated holistic approach is very much necessary to provide higher potency against microbial spoilage, keeping in mind the safe limits of each treatment.

KEYWORDS

- **bacteria**
- **fungi**
- **pathogenic microorganisms**
- **post-harvest diseases**
- **viruses**

REFERENCES

Abdellatif, E., Kaluzna, M., Helali, F., Cherif, M., Janse, J. D., & Rhouma, A., (2015). First report of citrus bacterial blast and citrus black pit caused by *Pseudomonas syringe* pv. *syringae* in Tunisia. *New Disease Reports, 32*, 35.

Akhtar, K. P., Shakir, A. S., & Sahi, S. T., (1999). Physiological studies of fungi causing postharvest losses of tomato fruits. *Pakistan Journal of Phytopathology, 11*, 25–29.

Banik, A. K., Kaiser, S. A. K. M., & Dhua, R. S., (1998). Influence of temperature and humidity on the growth and development of *Colletotrichum gloeosporioides* and *Diplodia natalensis*, causing postharvest fruit rot of mango. *Advances in Plant Sciences, 11*, 51–57.

Barkai-Golan, R. (2001a). Chapter 2: Postharvest disease initiation (pp. 3–24). In: *Postharvest diseases of fruits and vegetables: development and control*. Elsevier. https://doi.org/https://doi.org/10.1016/B978-044450584-2/50002-2.

Barkai-Golan, R. (2001b). Chapter 5: Attach mechanisms of the pathogens (pp. 54–65). In: *Postharvest diseases of fruits and vegetables: development and control.* Elsevier. https://doi.org/https://doi.org/10.1016/B978-044450584-2/50005-8.

Barkai-Golan, R. (2001c). Chapter 4: Factors Affecting Disease Development (pp. 33–53). In: *Postharvest diseases of fruits and vegetables: development and control.* Elsevier. https://doi.org/https://doi.org/10.1016/B978-044450584-2/50004-6.

Barkai-Golan, R. (2001d). Chapter 3: Each Fruit or Vegetable and its Characteristic Pathogens (pp. 25–32). In: *Postharvest diseases of fruits and vegetables: development and control.* Elsevier. https://doi.org/https://doi.org/10.1016/B978-044450584-2/50003-4.

Barkai-Golan, R. (2001e). Chapter 6: Host Protection and Defense Mechanisms (pp. 66–93). In: *Postharvest diseases of fruits and vegetables: development and control.* Elsevier. https://doi.org/https://doi.org/10.1016/B978-044450584-2/50006-X.

Biruma, M., Pillay, M., Tripathi, L., Blomme, G., Abele, S., Mwangi, M., & Turyagyenda, L., (2007). Banana Xanthomonas wilt: A review of the disease, management strategies, and future research directions. *African Journal of Biotechnology*, *6*(8).

Broadbent, L., (1965). The epidemiology of tomato mosaic: XI. Seed-transmission of TMV. *Annals of Applied Biology*, *56*(2), 177–205.

Coates, L., & Johnson, G., (1997). Postharvest diseases of fruit and vegetables. *Plant Pathogens and Diseases*, 533–547.

Collum, T. D., Stone, A. L., Sherman, D. J., Rogers, E. E., Dardick, C., & Culver, J. N., (2019). Translatome profiling of plum poxvirus infected leaves in European plum reveals temporal and spatial coordination of defense responses in phloem tissues. *Molecular Plant-Microbe Interactions*, (ja).

Coplin, D. L., (1980). *Erwinia carotovora* var. *crotovora* on bell peppers in Ohio. *Plant Disease Reporter, 64*, 191–194.

Coskuntuna, A., & Ozer, N., (2008). Biological control of onion basal rot disease using *Trichoderma harzianum* and induction of antifungal compounds in onion set following seed treatment. *Crop Protection, 27*(3–5), 330–336.

Da Rocha, N. A. C., Beaudry, R., Maraschin, M., Di Piero, R. M., & Almenar, E., (2019). Double-bottom antimicrobial packaging for apple shelf-life extension. *Food Chemistry*, *279*, 379–388.

Dasgupta, M. K., & Mandal, N. C., (1989). *Postharvest Pathology of Perishables.* Oxford and IBH Publishing Co. Pvt. Ltd., New Delhi., India.

De León, L., Siverio, F., López, M. M., & Rodríguez, A., (2011). *Clavibacter michiganesis* subsp. michiganensis, a seedborne tomato pathogen: Healthy seeds are still the goal. *Plant Disease*, *95*(11), 1328–1338.

Diab, S., Henis, Y., Okon, Y., & Bashan, Y., (1982). Effects of relative humidity on bacterial scab caused by {Xanthomonas} campestris pv. vesicatoria on pepper. *Phytopathology*, *72*(9), 1257–1260.

Duffus, J. E., Liu, H. Y., Wisler, G. C., & Li, R., (1996). Lettuce chlorosis virus: A new whitefly-transmitted clostero virus. *European Journal of Plant Pathology*, *102*(6), 591–596.

Eckert, J. W., & Ogawa, J. M., (1988). The chemical control of postharvest diseases: Deciduous fruits, berries, vegetables, and root/tuber crops. *Annual Review of Phytopathology*, *26*(1), 433–469.

Eckert, J. W., (1978). Pathological diseases of fresh fruits and vegetables. In: Hultin, H. O., & Milner, N., (eds.), *Postharvest Biology and Biotechnology* (pp. 161–209). Food and Nutrition Press, Westport.

Eden, M. A., Hill, R. A., Beresford, R., & Stewart, A., (1996). The influence of inoculum concentration, relative humidity, and temperature on infection of greenhouse tomatoes by *Botrytis cinerea. Plant Pathol., 45*, 798–806.

Fabi, J. P., Broetto, S. G., Da Silva, S. L. G. L., Zhong, S., Lajolo, F. M., & Do Nascimento, J. R. O., (2014). Analysis of papaya cell wall-related genes during fruit ripening indicates a central role of polygalacturonases during pulp softening. *PLoS One*, *9*(8). https://doi.org/10.1371/journal.pone.0105685.

Food and Agriculture Organization of the United Nations, (2013). *Food Wastage Footprints-Impacts on Natural Resources.* Summary Report (2013). http://www.fao.org/news/story/en/item/196402/icode/#targetText=The%20global%20volume%20of%20food,into%20the%20atmosphere%20per%20year (accessed on 21 December 2020).

Francki, R. I. B., & Hatta, T., (1981). Tomato spotted wilt virus. In: Kurstakm, E., (ed.), *Hand Book of Plant Virus Infection and Comparative Diagnosis* (pp. 492–512). Elsevier/North-Holland Biomedical Press, Amsterdam.

Ghuffar, S., Irshad, G., Aslam, M. F., Naz, F., Mehmood, N., Hamzah, A. M., & Gleason, M. L., (2018). First report of mucor fragilis causing bunch rot of grapes in Punjab, Pakistan. *Plant Disease*, *102*(9), 1858.

Gupta, A. K., Choudhary, R., Bashyal, B. M., Rawat, K., Singh, D., & Solanki, I. S., (2019). First report of root and stem rot disease on papaya caused by *Fusarium falciforme* in India. *Plant Disease*, (ja).

Hasan, M. F., Islam, M. A., & Sikdar, B., (2018). Biological control of bacterial leaf spot disease of papaya (*Carica papaya*) through antagonistic approaches using medicinal plants extracts and soil bacteria. *Int. J. Pure App. Biosci.*, *6*(1), 1–11.

Hollier, C. A., & King, S. B., (1985). Effects of temperature and relative humidity on germinability and infectivity of *Puccinia polysora* uredospores. *Plant Disease (USA)*.

Hussain, A. Z. H. A. R., Khan, S. W., Awan, M. S., Ali, S., Abbas, Q. A. M. M. A. R., Ali, Z., & Ali, S. H., (2017). Potato black scurf, production practices and fungi toxic efficacy of *Rhizoctonia Solani* isolates in hilly areas of Gilgit-Baltistan Pakistan. *Pak. J. Bot.*, *49*(4), 1553–1560.

Janisiewicz, W. J., & Korsten, L., (2002). Biological control of postharvest diseases of fruits. *Annual Review of Phytopathology*, *40*(1), 411–441.

Jarvis, W. R., (1994). Latent infections in the pre-and postharvest environment. *Hort. Science.*, *29*(7), 749–751. Retrieved from: http://hortsci.ashspublications.org/content/29/7/749.full.pdf (accessed on 21 December 2020).

Kamble, A. K., (2017). *In vitro* efficacy of different chemicals and biological agents against *Xanthomonas campestris* pv. viticola causing bacterial leaf spot of grapes. *International Journal of Agriculture Sciences,* ISSN: 0975-3710.

Kao, J., Jia, L., Tian, T., Rubio, L., & Falk, B. W., (2000). First report of Cucurbit yellow stunting disorder virus (genus Crinivirus) in North America. *Plant Disease*, *84*(1), p. 101.

Kawamoto, S., & Lorbeer, J. W., (1974). Infection of onion leaves by *Pseudomonas cepacia*. PDF. *Phytoparasitica*, *64*, 1440–1445. https://doi.org/10.1094/Phyto-64-1440.

Lapwood, D. H., (1977). Factors affecting the field infection of potato tubers of different cultivars by blight *(Phytophthora infestans)*. *Ann. Appl. Biology, 85*, 23–42.

Legard, D. E., Xiao, C. L., Mertely, J. C., & Chandler, C. K., (2000). Effects of plant spacing and cultivar on incidence of botrytis fruit rot in annual strawberry. *Plant Disease*, *84*(5), 531–538. https://doi.org/10.1094/PDIS.2000.84.5.531.

Li, B. J., Li, P. L., Li, J., Chai, A. L., Shi, Y. X., & Xie, X. W., (2017). First report of fusarium root rot of *Solanum melongena* caused by *Fusarium solani* in China. *Plant Disease*, *101*(11), 1956–1956.

Lin, C. P., Tsai, J. N., Ann, P. J., Chang, J. T., & Chen, P. R., (2017). First report of Rhizopus rot of strawberry fruit caused by *Rhizopus stolonifer* in Taiwan. *Plant Disease*, *101*(1), 254.

Mandal, N. C., (1981). *Postharvest Diseases of Fruits and Vegetables.* Doctoral Thesis, Visva-Bharati University, India.

Manicom, B. Q., (1986). Factors affecting bacterial black spot of mangoes caused by *Xanthomonas campestris* pv. *mangiferaeindicae*. *Annals of Applied Biology*, *109*(1), 129–135.

Martin, R. R., Tzanetakis, I. E., Gergerich, R., Fernandez, G., & Pesic, Z., (2003). Blackberry yellow vein associated virus: A new crinivirus found in blackberry. In: *X International Symposium on Small Fruit Virus Diseases* (Vol. 656, pp. 137–142).

Mensi, I., Jabnoun-Khiareddine, H., Zarrougui, N. E., Zahra, H. B., Cesbron, S., Jacques, M. A., & Daami-Remadi, M., (2018). First report of tomato bacterial speck caused by *Pseudomonas syringae* pv. *tomato* in Tunisia. *New Disease Reports, 38.*

Molina, J. J., & Harrison, M. D., (1980). The role of *Erwinia carotovora* in the epidemiology of potato blackleg. II. The effect of soil temperature on disease severity. *American Potato Journal, 57*(8), 351–363.

Moriones, E., Praveen, S., & Chakraborty, S., (2017). Tomato leaf curl New Delhi virus: An emerging virus complex threatening vegetable and fiber crops. *Viruses, 9*(10), 264.

Myung, I. S., Lee, J. Y., Yun, M. J., Lee, Y. H., Lee, Y. K., Park, D. H., & Oh, C. S., (2016). Fire blight of apple, caused by *Erwinia amylovora*, a new disease in Korea. *Plant Disease, 100*(8), 1774–1774.

Nabi, S. U., Raja, W. H., Kumawat, K. L., Mir, J. I., Sharma, O. C., & Singh, D. B., (2017). *Post-Harvest Diseases of Temperate Fruits and Their Management Strategies: A Review, 5*(3), 885–898.

Narayanasamy, P., (2005). *Postharvest Pathogens and Disease Management.* https://doi.org/10.1002/0471751987.

Ojaghian, S., Wang, L., Xie, G. L., & Zhang, J. Z., (2019). Inhibitory efficacy of different essential oils against storage carrot rot with antifungal and resistance-inducing potential. *Journal of Phytopathology, 167*(9), 490–500.

Pallavi, R., Uma, T., & Nitin, D., (2014). Postharvest fungal diseases of fruits and vegetables in Nagpur. *Int. J. Life Sciences,* (A2), 56–58.

Pearson, R. D., & Hall, D. H., (1975). Factors affecting the occurrence and severity of black mold in ripe tomato fruit caused by *Alternaria alternata. Phytopathology, 65*, 1352–1359.

Perombelon, M. C. M., & Salmond, G. P. C., (1995). Bacterial soft rots. In: Singh, U. S., & Kohmoto, K., (eds.), *Pathogenesis and Host Specificity in Plant Disease* (Vol. 1, pp. 1–20). Pergamon Press Ltd., Oxford, U.K.

Phipps, R. H., & Park, J. R., (2002). Environmental benefits of genetically modified crops: Global and European perspectives on their ability to reduce pesticide use. *Journal of Animal and Feed Sciences, 11*(1), 1–18.

Romanazzi, G., Smilanick, J. L., Feliziani, E., & Droby, S., (2016). Integrated management of postharvest gray mold on fruit crops. *Postharvest Biology and Technology, 113*, 69–76.

Rotem, J., Wooding, B., & Aylor, D. E., (1985). The role of solar radiation, especially ultraviolet, in the mortality of fungal spores. *Phytopathology, 75*(5), 510–514.

Rushing, J. W., Keinath, A. P., & Cook, W. P., (1999). Postharvest development and transmission of watermelon fruit blotch. *HortTechnology, 9*(2), 217–219.

Saltveit, M. E., (1999). Effect of ethylene on quality of fresh fruits and vegetables. *Postharvest Biology and Technology, 15*(3), 279–292.

Segev, L., Wintermantel, W. M., Polston, J. E., & Lapidot, M., (2004). First report of *Tomato chlorosis* virus in Israel. *Plant Disease, 88*(10), 1160.

Sharpies, R. O., (1984). The influence of pre-harvest conditions on the quality of stored fruit. *Acta Hortic., 157*, 93–104.

Silveira, N. S. S., Michereff, S. J., Mariano, R. D. L. R., Noronha, M. A., & Pedrosa, R. A., (2000). Virulence and dispersion gradient of fungi causing postharvest rots in tomato fruits. *Summa Phytopathologica, 26*, 422–428.

Singh, D., & Sharma, R. R., (2018). Postharvest diseases of fruits and vegetables and their management. In: *Postharvest Disinfection of Fruits and Vegetables* (pp. 1–52). Academic Press.

Snowdon, A. L., (1990). *Post-Harvest Diseases and Disorders of Fruits and Vegetables* (Vol. 1, p. 302). General Introduction and Fruits, CRC Press, Inc., Boca Raton.

Soylu, E. M., & Kose, F., (2015). Antifungal activities of essential oils against citrus black rot disease agent *Alternaria alternata. Journal of Essential Oil-Bearing Plants, 18*(4), 894–903.

Templeton, M. D., Warren, B. A., Andersen, M. T., Rikkerink, E. H., & Fineran, P. C., (2015). Complete DNA sequence of *Pseudomonas syringae* pv. *actinidiae*, the causal agent of kiwifruit canker disease. *Genome Announc.*, *3*(5), e01054–15.

Tian, S., (2007). *Management of Postharvest Diseases* (pp. 55–71).

Tzanetakis, I. E., Halgren, A. B., Keller, K. E., Hokanson, S. C., Maas, J. L., McCarthy, P. L., & Martin, R. R., (2004). Identification and detection of a virus associated with strawberry pallidosis disease. *Plant Disease*, *88*(4), 383–390.

Van, D. W., Van, J. M., & Van, B. J., (2004). Factors affecting survival of *Clavibacter michiganensis* subsp. *sepedonicus* in water. *Journal of Phytopathology*, *152*(3), 161–168.

Verhoeff, K., & Liem, J. I., (1975). Toxicity of tomatine to *Botrytis cinerea*, in relation to latency. *Journal of Phytopathology*, *82*(4), 333–338.

Verhoeff, K., (1970). Spotting of tomato fruits caused by *Botrytis cinerea*. *Netherlands Journal of Plant Pathology, 76*, 219–226.

Wintermantel, W. M., (2004). Emergence of greenhouse whitefly (*Trialeurodes vaporariorum*) transmitted criniviruses as threats to vegetable and fruit production in North America. *APSNet Feature Story.* Available at: Http://www.apsnet.org/publications/apsnetfeatures/Pages/GreenhouseWhitefly.aspx (accessed on 21 December 2020).

World Bank, World Development Indicators, (2018). *Population Total*. Retrieved from: https://databank.worldbank.org/reports.aspx?source=2&series=SP.POP.TOTL&country=#selectedDimension_WDI_Time (accessed on 21 December 2020).

Xu, X. M., Guerin, L., & Robinson, J. D., (2001). Effects of temperature and relative humidity on conidial germination and viability, colonization and sporulation of *Monilinia fructigena*. *Plant Pathology*, *50*(5), 561–565.

INDEX

F

L

M

N

O

Q

R